The Urban Book Series

Aims and Scope

The Urban Book Series is a resource for urban studies and geography research worldwide. It provides a unique and innovative resource for the latest developments in the field, nurturing a comprehensive and encompassing publication venue for urban studies, urban geography, planning and regional development.

The series publishes peer-reviewed volumes related to urbanization, sustainability, urban environments, sustainable urbanism, governance, globalization, urban and sustainable development, spatial and area studies, urban management, urban infrastructure, urban dynamics, green cities and urban landscapes. It also invites research which documents urbanization processes and urban dynamics on a national, regional and local level, welcoming case studies, as well as comparative and applied research.

The series will appeal to urbanists, geographers, planners, engineers, architects, policy makers, and to all of those interested in a wide-ranging overview of contemporary urban studies and innovations in the field. It accepts monographs, edited volumes and textbooks.

More information about this series at http://www.springer.com/series/14773

Laxmi Ramasubramanian • Jochen Albrecht

Essential Methods for Planning Practitioners

Skills and Techniques for Data Analysis, Visualization, and Communication

Foreword by Mike Batty

Springer

Laxmi Ramasubramanian
Department of Urban Policy and Planning
Hunter College
New York, NY, USA

Jochen Albrecht
Department of Geography
Hunter College
New York, NY, USA

ISSN 2365-757X ISSN 2365-7588 (electronic)
The Urban Book Series
ISBN 978-3-319-68040-8 ISBN 978-3-319-68041-5 (eBook)
DOI 10.1007/978-3-319-68041-5

Library of Congress Control Number: 2017954945

Printed on acid-free paper

This Springer imprint is published by Springer Nature
The registered company is Springer International Publishing AG
The registered company address is: Gewerbestrasse 11, 6330 Cham, Switzerland

Foreword

Just over one hundred years ago, the great American architect and city planner, Daniel Burnham, said: "Make no little plans; they have no magic to stir men's blood and probably themselves will not be realized. Make big plans; aim high in hope and work, remembering that a noble, logical diagram once recorded will never die ..." His phrase has been quoted and re-quoted many times throughout the last century but slowly and surely, the idea of big plans has fallen off our agenda. When you examine the kinds of plans that practitioners routinely work with now, you realize that they are much more modest than the grandiose master plans of the last century. There are many, many small plans compared to big ones. Many are individual projects that planners then have to be able to integrate into the wider scene, and they often consist of small-scale changes, sometimes leading to much bigger changes, that have to be anticipated by the planner. To an extent, this has always been the case, but as we have begun to realize the complexity of the planning task before us, we have begun to focus our attention on finer and finer details right down to the most basic elements in the community. This has re-orientated our field to the local, the small scale, and the individual.

This is in my view entirely what the focus of planning should be about. We must work with the small scale, for that is where urban change has the most impact on peoples' lives and on their quality of life. Moreover, this is where we can engage best with those whose lives are most affected by the pressures for urban change, by the imposition of plans, and by the inevitable conflicts that occur over the use and allocation of scarce resources, particularly land. What Dr. Ramasubramanian and Dr. Albrecht do in this book is to pose questions as to how one might best use and apply the many tools and techniques available for planning preparation and community participation in the planning process that have been developed over the last fifty years. They articulate how we might best embed these in processes of community and citizen engagement that infuse the search for good plans with the most useful ways of researching and communicating these ideas to a wider constituency.

This is not a book that is fashioned as a step-by-step account of how these techniques and tools are structured. It is not a technical book, nor is it a manual for enabling the reader to construct techniques from scratch. It is a book that takes the planning practitioner to be the heart of the process of planning and to inform those processes with the kinds of methods that will help the professional engage best with his or her wider constituency in the search for the best plan. This is a very brave way of introducing planning methods. For many years, there has been a key schism between those developing methods and those seeking to use them, and it is rare to find good discussions of the perils of such integration. Even rarer are examples that actually demonstrate this. This book advocates a point of view rather than a manifesto for action, and it argues that the best way of integrating methods into planning is through active planning applications that engage the community and imbue the citizenry with the power to use these new tools and develop them for their own specific ends.

The authors provide what they term a "ready-to-use guidebook" based on a "customized and curated compendium of methods and techniques". This compendium can be used as a handy reference source for a series of tools and techniques that almost act by way of a checklist, a kind of backcloth for a wide range of community planning projects. To tell their story, however, they identify three key issues that most of us, if not all of us, would agree to be the most important issues of the twenty-first century: namely, urbanization, demographic shifts, and climate change. In essence, the world's population may stabilize this coming century, and certainly overall growth will fall while at the same time the inexorable drift to cities will continue. By the end of this century, the world will be largely urbanized, and the consequences for planning are thus enormous. Demographic change, of course, will be confounded by an aging society with substantial advances in medicine, and life expectancy will be prolonged as much by surgical intervention as by diet, lifestyle, and the elimination of disease using pharmaceuticals. All of this will be set against a background of climate change, and as more than half the world's cities lie in coastal areas, sea-level rise will be a major issue. The impacts of these key forces on urban sprawl, smart growth, diversity in cities, environmental quality, issues of resilience, and on how communities will participate in the processes designed to tackle these major issues can best be handled using the tools and methods that the authors identify and demonstrate in the various chapters of this book. Again, all this is set against a background of continuing technological change that is foisting a digital revolution on the way we will live in cities during this century and beyond.

As Dr. Ramasubramanian and Dr. Albrecht argue, "planning can only be successful if it is adapted to the situational context", and they develop this theme early in their exposition using two case studies from New York City. The first is a small area of some 2–3 square kms in the South Bronx at Hunts Point. This is a very mixed low-income and de-industrialized community that shows all the scars of contemporary big city living where poverty is never far away, where the local environment is polluted and dirty, and where access to transportation is not as good as most other places in the metropolis. Their second case study is Roosevelt Island which has quite different problems. It is richer and is being gentrified quite rapidly as well as being a recipient for new high-tech industries and science research centers in the City. These case studies set the context for the introduction of methods that are outlined after the case studies have identified key issues in terms of urban change that the planning processes assumed here are focused upon. The way the authors continually refer to them in guiding their use of methods in practice is extremely effective.

It is worth making a distinction between the somewhat heavier, more scientific explanatory methods useful for planning and the lighter, more future-based tools that are much more appropriate for planning processes that involve wide and deep dialogues with the affected communities. It is these future-based methods such as Delphi, the Futures Wheel, and Forecasting methods such as those in transportation planning that the authors argue are key to those that practitioners should invoke in their analysis and scenario writing. It is here that the more local focus reasserts itself and the development of a variety of crowdsourcing, sensing, behavioral mapping, participant observation, ethnographic analyses, including GIS, spatial analytical tools and new approaches such as geo-design, are central to these processes.

When these methods are considered collectively, then the focus on planning and civic engagement with such methods being key to this is developed. To an extent, the entire book is orientated to this kind of engagement, which moves beyond public participation per se to processes that "support and inform …. day-to-day work" of the planning professionals and their involved communities. The rationale for the way the authors develop their thesis becomes apparent as they develop this argument. They argue "… that all planners have a responsibility, an obligation and the skills to support and nurture civic engagement…" and in this, they conclude that planning expertise is as much a part of the local community as it is a part of the training and professional skills of the planner. In this, planners are agents of change, they are part of the transformational process that turns the present into the future, and the logic of this book is that the tools and methods identified are necessary but also subservient to the wider dictates of the community development process. What is worth taking from this book is that these transformational processes must be part of a dialogue between planners and their communities. This, of course, is being massively enhanced by the new digital world of data and participation that is based on the idea of "digital storytelling", an idea that they discuss throughout their book.

The notions of planning as dialogue, as mediation, as storytelling, as agenda setting, and as turning knowledge into action are all key to the way Dr. Ramasubramanian and Dr. Albrecht develop their argument. This is an innovative and unusual way of introducing methods and it is convincing in that it weaves the notions of a methodologically explicit form of planning into a context which is highly applicable to smaller scale, intensive kinds of projects that now characterize planning in many places around the globe. They provide key messages for how we should develop planning in a future consistent with the digital world we have now entered. The messages in this book are important and relevant to the physical and social development of our cities in the twenty-first century.

Read on and enjoy!

University College London, London, UK Michael Batty

Preface

This is a book for planning practitioners – for those aspiring to become planners, new graduates, practitioners changing careers, and anyone who is interested in understanding what planners do. This is also a book about planning methods and techniques. As coauthors, we have expertise and experience in architecture, geography, urban planning, and Geographic Information Science; we set out to write a book that organizes planning methods and techniques within a theoretical context and describe the use of the methods in the context of undertaking conventional planning activities.

Planning practitioners all over the world, particularly those working for local governments, encounter complex challenges in their everyday work. They combat a weary societal cynicism that dismisses planning as ineffective or irrelevant while simultaneously chafing at perceived overreach that undermines self-determination. Planning offices are under-resourced and planners often struggle as they strive to speak truth to power. Nevertheless, they persist!

We have both been fortunate to have worked with talented planning practitioners who demonstrated how to craft powerful and engaging narratives to capture the hearts and minds of different stakeholders, stories that wove a tapestry linking the experiential knowledge of diverse stakeholders with appropriate analysis and data-driven evidence to create transformational change. These successful practitioners have honed their craft over time, learning how to exercise practical judgment to solve complex problems.

For recent graduates and newly employed planners, especially for women and people of color, understanding and practicing the craft is not easy. There is seldom time to reflect about why and how certain actions and decisions were taken and why certain methods were used – much is lost in the everyday urgency to get work completed. At the same time, new planners are more likely to get siloed, working on one aspect of planning, and not get to experience the big picture. Our book provides some guidance to ease some of these anxieties. It also challenges planners to think differently about their work.

We are grateful to our friends and family from around the world who accepted that we were unavailable to them while we worked on this project. They sent support from afar and accepted our strange preoccupation with good grace. We also wish to thank our professional colleagues and our students who have helped to sharpen our thinking about many of the issues we discuss in the book. We are grateful to Judy Colby-George and Anna Slatinsky who read and commented on the draft and provided us with useful feedback.

We were truly lucky to be able to work with a talented architect-planner Mr. Marco Castro in organizing the graphics for this book. What began as a routine task to create maps and drawings transformed into an interesting and engaging collaboration about data visualization, cartography, and information communication. As we develop a digital presence for this book, we are excited to continue our collaboration with him. Look for us online at *allthingsplanning.org.*

As professors engaged in the business of preparing practitioners, we constantly balance our desire to retreat into the wonky and analytical world of academic scholarship with our urge to solve practical problems. This book is our way of achieving that balance – bringing theory to practitioners to encourage a more reflective and politically engaged practice. We wrote this book because we care deeply about the field and the profession. In our view, the field and profession can be strong only when its practitioners feel empowered. We sincerely hope that familiarity with the methods and techniques discussed in the book supports that process.

New York, NY, USA

Laxmi Ramasubramanian
Jochen Albrecht

Contents

List of Figures

List of Tables

Chapter 1
Planning as Storytelling

1.1 Introduction

Planning, in essence, is a set of future-oriented activities where individuals and groups organize their current understandings of their circumstances to be better prepared for the future (Wildavsky 1973; Alexander 1981). Whether we are aware of it or not, all of us plan constantly. Planning is not a set of activities that is related to one's chronological age or activities. We could say that planning is hardwired into our psyche. Yet, in the United States, there is considerable skepticism and wariness about the role of planning in/for the public realm; many people believe that planning infringes on individual property rights while others believe that planning creates new and onerous regulations that inhibit growth and development. In resource-poor communities, planning is viewed as an enterprise that only serves the interests of the wealthy and powerful. When elected officials engage in conversations about planning, they are often feted by some sections of the electorate while being simultaneously vilified by others. It is this challenging context for the professional planner at the beginning of her career that motivates us to write this book.

Our society's ambivalence toward planning places limits on the kinds of projects, programs, and policies that local governments can pursue. In the United States, the ambitious and visionary planning projects that were popular in the late nineteenth and early twentieth centuries now seem like fantasies. As a matter of fact, one of our case study examples in Chapter 3, Roosevelt Island, may be one of the last examples for such grand plans. Local government planning has become a tedious, time-consuming, bureaucratic, and regulatory process. Innovation and creativity are sometimes stifled by budgetary realities or to avoid litigation. Daniel Burnham's extortion *to make big plans* is often ignored. Nevertheless, planners are engaged in shaping our built environment in myriad ways; our streets, our parks and public

L. Ramasubramanian, J. Albrecht, *Essential Methods for Planning Practitioners*, The Urban Book Series, https://doi.org/10.1007/978-3-319-68041-5_1

places, our transit stations, and our airports are in a process of transformation to address present-day urban problems and challenges.

1.2 Scope and Purpose

In this book, we celebrate the integrative and syncretic qualities of planning practice. This is brought home by reserving a whole chapter on two case studies, where we provide an in-depth description of the situational complexities of real-world planning projects. The everyday practice of planning, in our view, is a craft. Like all crafts, it is honed over time. Good planners are not technocrats, although they use their technical expertise to make compelling arguments; good planners are not demagogues, although they create coherent narratives to convince even the most hostile naysayers; they are not politicians, although they are persuasive and convincing in advocating particular courses of action. Good planners work with the public, viewing them as allies in their efforts to solve complex problems. In this book, we put forward a visionary and perhaps radical approach, a way to reimagine how the field and the profession can engage with citizens from all walks of life, embracing diversity and complexity as part of the process and in the creation of the final product. We suggest that planners and citizens can work together, collaboratively, creatively, and proactively, focusing on problem solving rather than creating distinct spheres of engagement, territories, and spaces where disagreements are played out.

In our work as planning educators and practitioners for over 20 years, we have found that good planners are able to construct accurate and dynamic socio-spatial narratives that provide some rich understanding of places and their experience. In our book, we describe tools, methods, and techniques that will enable aspiring planners and planning professionals to become better at their craft – in other words, to become better planners.

Our book, "Essential Methods for Planning Practitioners" is written for individuals who are planning to enter the planning profession, as graduate students or those who are in search of that all-important first planning job, armed with a graduate degree in planning or a related field and some internship experience. The book can be a very useful and handy desk reference for anyone who is working in business, government, or nonprofit sector, undertaking the many different types of work associated with "doing planning." A junior planner can be called upon to review and synthesize relevant literature, design and conduct a survey, develop a community engagement plan, analyze data from a variety of sources, create maps and other spatial analyses to support particular policy positions, and/or manage a project. Many of these topics have received book-length treatment and are often covered in one or more required "methods" classes that every student must take during their academic career. As scholars and educators, we have read many of these books and assigned them as readings in our graduate classes. In this book, we take a different approach – we discuss the applications of different methods in their social and institutional contexts with the goal of future-oriented problem solving.

1.3 What Do Planners Do?

The role of the professional planner, particularly in the United States, has changed quite dramatically over the last 50 years, as exemplified in Table 3.1, where we juxtapose the five consecutive planning paradigms of our case study areas. Contemporary urban planners confront problems across many spatial scales (see, e.g., the conflict between the regional and local interests in our Hunts Point case study), addressing hyper-local neighborhood issues as well as city-wide and regional concerns. As the field of planning has become more complex, a high degree of specialization and differentiation has occurred. Planners now work alongside other professionals such as architects, cartographers, demographers, engineers, landscape architects, urban designers, statisticians, public health and social media experts. In this context, how and why should planning professionals engage in participatory planning and design in the twenty-first century? In the last 50 years, planning and design professionals have accepted the idea that the public must be consulted about important decisions affecting their neighborhoods and communities. However, the methods and approaches used to engage the public and communicate with them remain rooted in the fractious 1960s. Our book assembles and organizes a selected range of methods and techniques that every planning practitioner should know. Our book is unique because it is not a methods textbook or even a reference book but one that links different aspects of the planning/policymaking enterprise with the appropriate methods and approaches – thus contextualizing the use of specific methods and techniques within a sociopolitical and ethical framing. Planners, especially those who are on the front lines, often feel anxious and underprepared for the demands of their job. As new entrants into the profession, they are often confronted with a data deluge (see the range of datasets underlying our case studies that can be found on the website accompanying our book, allthingsplanning.org) and a database of methods and techniques without specific guidance that helps them assess the value of effectiveness of one over another. At the same time, because technologies, data, and consequently analytical techniques tend to change rapidly, it is better to provide practitioners with a guide of how to select and deploy methods and techniques, rather than attempt to provide an inventory which would become obsolete quickly. These challenges are well known to seasoned leaders and managers; namely, the context (the nature of the problem) should determine the choice of appropriate methods and techniques. At the same time, practical considerations, like available expertise, allocated budget, and time available, should also reasonably influence these choices.

In the present milieu, there is often some confusion about the role or roles that planners are expected to play: Are planners responsible for spatial planning and land use? Are they the experts who ensure compliance with zoning laws and height restrictions? Are planners facilitators who mediate between experts and everyday citizens? Do planners address quality-of-life issues related to traffic, air quality, and noise? Are planners technicians? Do they set policy? As we will illuminate in the following chapter, the answer is yes to all of the above, and it is this diversity and breadth that makes a planner's job exciting, interesting, and sometimes frustrating (Hoch 2011). According to the American Planning Association, "professional planners help create a broad vision for the community. They also research, design, and develop programs; lead public processes; effect social change; perform technical analyses; manage; and educate. Some planners focus on just some of these roles, such as transportation planning, but most will work at many kinds of planning throughout their careers" (American Planning Association 2017). As future-oriented and pragmatic decision-makers, planners are required to concern themselves with larger concerns such as public health, natural resource management, and climate change, phenomena that, as our case studies exemplify, have socio-spatial consequences in the neighborhoods and communities where they work. Planners often find that the problem they are trying to address has its origins within a different locus of authority, complicating and confounding traditional decision-making processes. The geographic scale at which planners work continues to be dynamic. The institutional contexts within which they work are as varied.

1.4 Future-Oriented Problem Solving: The Climate Change Imbroglio

Let's briefly examine the complex issue of climate change. That the earth's climate is changing is beyond doubt. There is a scientific consensus that human actions are one of the major drivers of this change and that human activities, particularly in the last 100 years or so, have exacerbated the situation (IPCC 2001). Data and evidence are all around us, in the scientific literature and the popular press. However, even after setting aside the minority view that denies any notion of climate change, there is a surprising lack of consensus about scale, scope, and impacts of this phenomenon. Bench scientists, who have the luxury to do so, advocate for additional research (Wilby and Wigley 1997). They are often unable to issue precise or specific guidance about how to advise the public in everyday situations – can I build or, better yet, should I build in this location? Scientists are also unwilling to advise elected officials about the ethics of relocating an entire residential community permanently to protect against future deleterious climate change impacts such as sea-level rise (Stern and Taylor 2007). They prefer to provide the data to end users who must interpret and use that data to make difficult decisions.

In the meantime, alarmists, as well as anxiety-prone activists, criticize international treaties and agreements related to climate mitigation. These groups sometimes consider the processes of setting emissions targets through a political consensus process as privileging politics over science. Activist groups are more likely to call for government interventions in the form of regulations and mandates. They propose that governments should use a heavy hand and regulate the behavior and actions of present generations to protect the planet for future generations (Gillard et al. 2016). The hardships such regulations may impose on present generations are outweighed by the long-term environmental benefits to humankind.

The public, as individuals, is often paralyzed by the volume of information that comes their way, and they tend to ignore most of it (NORC 2016). A variety of interest groups – research centers, advocacy organizations, and self-help groups – mediate the relationship between these two extreme positions. Depending on the specialized focus of each group, they put forward, discuss, and debate a range of policy positions and highlight different aspects of the climate change challenge. One group might highlight the concerns of homeowners while another might emphasize the challenges of renters; still others might discuss the impacts of climate change mitigation/adaptation measures on economic prosperity and so on. In this context, planners serve as educators and facilitators – explaining the short- and long-term consequences of human-induced climate change to multiple publics (Ramasubramanian 2016). They also provide guidance (outside of regulations and mandates) about how to balance between and among competing interests to create a viable set of actions to prepare and plan for climate change impacts.

Climate change mitigation and adaption planning cannot be standardized across the globe. A climate change plan for a city or region is meaningless if it cannot accommodate the particularities of geography, the type and quality of the transportation infrastructure and housing stock, as well as the area's population density (Otto-Zimmermann 2011). Planners thus have to acquire additional expertise in different subjects to make meaningful and viable conclusions for the community/region they serve (Campbell-Lendrum and Corvalán 2007). Many additional factors are often considered to develop planning guidelines or more restrictive zoning ordinances. Accordingly, a planner's role is circumscribed within the institutional context where they choose to work, because planners seldom work in isolation as individual practitioners. Planners work in community-based organizations, as part of research groups, as educators, as activists, and as public servants, and across all levels of government (APA 2017). Government agencies are, by and large, the most significant employers of planning professionals (BLS 2016). As representatives of government, planners become responsible for translating the best scientific knowledge about human-induced climate change to a skeptical public. Planners develop reasonable and practical guidelines for mitigating and adapting to climate change to ensure stricter guidelines related to public safety (Showstack 2014). To fulfill their obligations, planners use a variety of analytical research methods. The information they assemble can be used to develop programs and projects. Planners become collectively responsible for protecting the *public interest*. In this context, the authors argue that planners have an additional and overarching mandate to protect the interest of those who cannot advocate for themselves, children, elderly, and other vulnerable populations as well as future generations who are not yet present (Hurlimann and March 2012).

Over the last 50 years, we have come to realize that planning is a transformative set of actions (Friedmann 1987) that require the involvement of a variety of types of expertise (Sarkissian et al. 2010). Many young professionals do not hold the title "planner," belying the actual work that they do (O*NET 2017). Furthermore, planners now seldom work alone – they work alongside elected officials, leaders of nongovernmental organizations, community activists, and members of the general public. Engagement with stakeholders and the public is now an essential ingredient in any planning process (Geertman and Stillwell 2009). Digital technologies have grown and evolved and now influence and impact every aspect of the planning enterprise (Batty et al. 2000; Couclelis 2004; Aurigi 2007; Mandarano et al. 2011). Contemporary planners, especially the ones who are successful, are attuned to the complexities of city and regional politics, particularly in how powerful interests can steer important conversations about social issues, and have had to become adept in using persuasion rather than enforcement authority to achieve change. Preparing urban planners to work in these challenging contexts is daunting and requires fresh educational and assessment approaches. This book will help prepare planners for a new and challenging role – as educators and guides in helping to create usable and useful knowledge that can be translated into implementable actions, in other words, to become digital storytellers.

1.5 Why This Book, Why Now?

As educators, we looked for a book to assign to our classes – but we could not find one despite our best efforts, so we decided to write one ourselves. Planning education is lagging behind the profession and has been slow to adapt to the changing environment of the workplace – planners, especially those in the early stages of their career, need to be familiar with a diverse range of methods, techniques, and skills and be prepared to deploy them thoughtfully by paying careful attention to the sociopolitical contexts within which they are deployed. Our book draws on tools, methods, and techniques from different but related academic disciplines, including behavioral geography, urban design, geographic information science, and public policy analysis.

Our book is a customized and curated compendium of methods and techniques – in other words, one book that will be on the bookshelf of every planning practitioner as a go-to methods guide for every planning practitioner and a starting point for reliable information and assistance in accomplishing their particular tasks. As longtime educators, familiar with the American higher education system, we have specific target audiences for this book. This book is intended for planning practitioners – our primary target audience is the newly minted planning graduate who is in the first 5 years of their career. These new practitioners are employed in all levels of government, in community-based organizations, and in the industry. In many instances, they have not yet cultivated relationships with supervisors, nor have they found mentors who will help them navigate life within their organization. Their responsibilities require them to have a wide range of expertise and to demonstrate great flexibility and ingenuity – they may be asked to design a survey during the first week, conduct a land-use survey in another week, or run GIS analyses during the third week. While each of them may have learned how to undertake each of these tasks during their student days, their work was conducted in a “safe” environment, with detailed guidance from the instructor. They probably had clearly defined objectives, clean data sets, and established analysis parameters. In a typical real-world context, such safety protections disappear, leaving many planners in a state of insecurity, which in turn limits their ability to act with confidence. Our ready-to-use guidebook provides curated content about methods and techniques to allow them to proceed with their work competently and confidently.

Both authors grew up with a generation of the Sage "little green books on research methods", which came in very handy when working on our dissertations or subsequently on research projects. These books inspired us to think about this particular book project. Rather than develop one little book per method or technique, we decided to build a compendium of methods curated and linked together by a narrative arc that addresses societal considerations related to access, equity, transparency, engagement, and accountability.

Presently, individuals with limited knowledge and understanding of planning principles or planning history are engaged in shaping the conversation about planning and policymaking (McArthur 2017). Often, these individuals are motivated with a commitment to their ideals. They advance a variety of causes such as mitigating climate change, creating transparency and accountability in government, or developing better transportation alternatives. We argue that these individuals will benefit from learning about a variety of research methods that will help inform and support their advocacy efforts. It may also limit inappropriate and nonproductive applications of statistical and mapping techniques, thereby improving the public policy discourse in our society.

1.6 Overview of Upcoming Chapters

Thus far, we have described our intended audience and discussed our motivation in writing this book. Our book is written for individuals entering the planning profession, graduate students, or those who are in search of that all-important first planning job, armed with a graduate degree in planning or a related field and some internship experience. We then outline our planning philosophy (planning for people), aiming to shatter the boundary line between the expert and the public. Our book assembles and organizes a selected range of methods and techniques that every planning practitioner should know, yet it is not a methods book in the traditional sense, where each is introduced in the abstract. Instead, we observe that the real-world planner always encounters interdependencies and that their recognition is crucial to successful planning. It is in this context that we then offer strategies, tactics, and modalities to the novice planner. The companion website allthingsplanning.org is intended to be filled with real-world examples (vignettes) and constantly updated to reflect the state of the art. Eventually, this website might develop a life of its own, where readers share experiences and contribute beyond what has been covered in this book. This chapter ends with an overview of the remaining parts of the book – basically the collection of the paragraphs that make up this annotated table of contents.

1.6.1 Chapter 2: Planning Challenges and the Challenges of Planning

The world of planning has drastically changed over the past few years. Understanding this change requires to briefly revisit the historic roots of planning and geography. Little has prepared the profession for the brave new data world though where data is confused with information (Cohen 2002; Kerski and Clark 2012). We juxtapose the current situation, where everybody thinks they are planners with traditional planning models, and prepare the reader to acknowledge people's experience while defending her expertise. In our conceptualization of planning, this is not a compromise position but a worldview. This chapter provides some answers to the questions that are often foremost in practitioners' minds – Why is any discussion of change so "political"? We argue that practitioners should understand the strengths and limits of planning (from the past) to plan more effectively in the future. In the second half of this chapter, we argue that each planning problem, regardless of whether it is simple or complex, large or small in scale, requires its own set of appropriate tools and recipes and illustrate this by outlining a framework of future planning challenges. In addition to climate change, the future long-term societal challenges we highlight are changing population characteristics, urbanization and migration, immigration, environmental quality and human health, and safety and security. The description of these tools and recipes forms the bulk of the remaining chapters.

1.6.2 Chapter 3: Case Studies

We would, however, fall into the same trap as traditional textbooks if we were to introduce these building blocks without context and, therefore, devote a whole chapter on two real-world planning situations (Roosevelt Island and Hunts Point, both in New York City). They follow a common workflow starting with a historic and geographic setting and a description of the people that live in or use the study area. This is followed by an analysis of the local needs and resources, which finally translate into planning challenges and opportunities. For the planning challenges, we follow the framework given in Chapter 2. Our case studies can be read by themselves as vignettes. Their main purpose, however, is to serve as a backdrop to the discussion of the other chapters, where we will make frequent references to the case studies. Throughout the book, we emphasize the need for a comprehensive and synthetic perspective; the methods described in Chapters 4 and 5, for instance, should not be applied out of context.

1.6.3 Chapter 4: Planning Grand

The first "methods" chapter discusses approaches that are related to the earliest stages of the planning process. As we stress the importance of communication, we start with the Delphi method as a precursor to e-democracy efforts (Rotondo 2012).

Traditionally, the domain of expert plan making and forecasting, the Delphi method can now be seen as bridging the gap between experts and non-experts. This naturally leads to a discussion of top-down vs. bottom-up approaches and the role of bias dealing with either community. While there still is a place for analog and face-to-face techniques, contemporary planning takes place in a new media landscape and with a citizenry of variable digital competency (ICTA 2005). In this context, the bulk of our envisioning techniques deals with larger and often incoherent groups. In our attempt to incorporate as many different perspectives as possible, we reflect on citizen science, crowdsourcing, and participatory mapping. Individual perspectives can be captured through digital storytelling and photovoice, both of which can be used in framing the research question as well as plan implementation (Chapter 7).

1.6.4 Chapter 5: Placemaking: Why Everything Is Local

Our fifth chapter can be considered our "data and needs assessment" chapter. Data, in all its diversity and messiness, helps us from making the mistake of relying completely on our instincts. Instincts can be good but can sometimes lead us astray. We deal with the myriads of ways we can get hold of the data that underlies every rational planning process. We juxtapose traditional (though sometimes altered in their character by the surprising variety of provenance) outsider perspectives like demographic profiles with various community-based techniques such as behavior maps, participant observation, sensor, perceptional mapping, or modern survey techniques. Regardless of how all this data is generated, it then needs to be critically assessed to be used in scenarios or simulations. We outline the opportunities afforded by these techniques but also provide cautionary notes on their constraints and reflect on the role of planners in a world where technically skilled but nonprofessional citizens are confounding officials with their data wizardry.

1.6.5 Chapter 6: Civic Engagement

This chapter argues that contemporary planning has to go beyond public outreach or even public participation and introduces civic engagement as the third phase after advocacy planning and citizen participation (Holman et al. 2007; Healey et al. 2008). We propose four principles of civic engagement that serve as measures of successful community involvement and place a range of techniques into a matrix of stakeholder involvement. Many of the strategies presented here have been field tested over the past decade, working with a range of special populations. We summarize this chapter with reflections on the interplay of techniques and the need for planners to creatively combine them as the situation demands.

Chapter 6 focuses exclusively on the complexities of civic engagement. Most planning projects or policies engage a range of state and non-state actors, and it is useful to acknowledge that all projects are highly reliant on successful engagement with a variety of stakeholders. Civic engagement can influence a project or program's success or failure – and in the twenty-first century, it is a truism to observe that good projects and ideas can sometimes be derailed because of poorly managed civic engagement processes (Malik and Wagle 2002).

Civic engagement is simultaneously a philosophy, a core value, an approach, and a set of methods that includes measurable indicators and outcomes (Hawkes 2001; Madera 2010; Gough 2015). It exemplifies the complexity of the planner's role in society, regardless of the particular job title or function she performs. For this reason, we treat civic engagement as a stand-alone chapter because the various methods discussed in Chapters 4 and 5 may have to be put through a civic engagement "test" in order to determine how these methods and approaches can and should be used within a planning process.

Note the use of the phrase "civic engagement" in this book rather than commonly used terms such as public outreach, public involvement, or public participation. These terms are used interchangeably in mainstream planning discourse, but they do not mean the same thing. In this book and chapter, the authors emphasize civic engagement – a term that connotes a genuine commitment to informed and reasoned deliberation and debate about a wide variety of planning issues over an extended time period. Shifting the conversations from public involvement to civic engagement is one of the true lessons a successful planning practitioner must learn in order to be effective in his or her job. It is impossible to be a leader in the field without mastering the art and science of civic engagement.

1.6.6 Chapter 7: Implementation and Sustainability

We emphasized in Chapter 4 the importance of communication and revisit the topic now from a different perspective. People make decisions. Decisions are shaped by ideologies and political agendas. This is a point we make early in the book. However, in this penultimate chapter, we want to emphasize the role of policy in driving good implementation and securing it for future generations, in other words making it sustainable. Therefore, we will discuss the interplay between planning and policy-making here, focusing on how to identify and assess different policy options. Once identified, they have to be formally accepted through legislation, regulation, or larger structural societal transformations. In our media-driven age, this means that we have to pull out all stops to assure project-internal communication as well as getting the word out via Web-based tools, meetups, and hackathons or engaging civic-minded people. We need to fund our plans, find allies, and, if necessary, create them by educating our clientele about modern planning practices and more generally how government works. Part of this includes not just the data fluency advocated for in Chapters 4 and 5 but also communication with data to reach different audiences and maintain a sustainable planning environment, as highlighted particularly in the Roosevelt Island case study. Arguably, this reaches into the realm of senior planning management, but it is never too early to learn about institutional constraints to turn them around into opportunities.

This chapter will allow planning students and practitioners to understand the differences between process-oriented policymaking, analysis-centric policymaking, and macroscale policymaking. It discusses how to identify and assess different policies and their intended effects on different population groups (equity considerations) as well as how to avoid unintended consequences. Often planning students, especially those trained in the technical fields like engineering and architecture, believe that their work ends when they present the technical studies and reports to their "bosses." The bosses are often elected officials who have earned the right to represent the public, whereas the planners or practitioners are considered unelected decision-makers. However, the planners/practitioners may likely be the individuals who provide continuity for projects and programs that are multi-year investments from the public purse. Constantly undertaking research or planning studies that do not result in tangible outcomes is costly in more ways than the obvious – it has the potential to undermine the public's trust in research.

1.6.7 Chapter 8: Epilogue

This concluding chapter provides a discussion about how the methods and techniques described in the book can collectively be used to create transformational and meaningful interventions. It begins with a discussion of individual skills that a planner ought to cultivate to be more effective in their chosen profession. Planners must hone their craft, like artists, but be attentive to the societal, institutional, and political frameworks that bound their work. The twenty-first-century planning must recognize the three major global planning challenges, urbanization, demography, and climate change, and how these challenges affect their everyday work. The chapter concludes by encouraging planners to create a balance between technical expertise, political *nous*, and ethical actions in order to create a more lasting and sustainable neighborhoods and communities. We realize that a book like this one is virtually outdated by the time it is published, which is why we will use the companion website allthingsplanning.org.

References

Alexander E (1981) If planning isn't everything, maybe it's something. Town Plan Rev 52(2):131–142

American Planning Association (APA) (2017) Where do planners work? Online resource, https://www.planning.org/onthejob/where.htm. Last accessed 5/7/2017

Aurigi A (2007) New technologies, same dilemmas: policy and design issues for the Augmented City. J Urban Technol 13(3):5–28. https://doi.org/10.1080/10630730601145989

Batty M, Chapman D, Evans S, Haklay M, Kueppers S, Shiode N, Smith A (2000) Visualizing the city: communicating urban design to planners and decision-makers, CASA working papers 32. Centre for Advanced Spatial Analysis. University College London (UCL), London

Bureau of Labor Statistics (BLS) (2016) Occupational employment and wages, May 2016, 19–3051 Urban and Regional Planners. Washington, DC: US Department of Labor. Online resource, https://www.bls.gov/oes/current/oes193051.htm. Last accessed 5/7/2017

Campbell-Lendrum D, Corvalán C (2007) J Urban Health 84 (Suppl 1):109–117. https://doi.org/10.1007/s11524-007-9170-x

Cohen B (2002) Community food security assessment toolkit, Economic research service. E-FAN-02-013. USDA Food Assistance & Nutrition Research Program, Washington, DC

Couclelis H (2004) The construction of the Digital City. Environ Plan B Urban Anal City Sci 31(1):5–19. https://doi.org/10.1068/b1299

Friedmann J (1987) Planning in the public domain: from knowledge to action. Princeton University Press, Princeton

Geertman S, Stillwell J (eds) (2009) Planning support systems best practice and new methods, Vol, vol 95. Dordrecht, Springer Science & Business Media

Gillard R, Gouldson A, Paavola J, Van Alstine J (2016) Transformational responses to climate change: beyond a systems perspective of social change in mitigation and adaptation. WIREs Clim Change 7:251–265. https://doi.org/10.1002/wcc.384

Gough M (2015) Reconciling livability and sustainability: conceptual and practical implications for planning. J Plan Educ Res 35(2):145–160. https://doi.org/10.1177/0739456X15570320

Hawkes J (2001) The fourth pillar of sustainability: culture's essential role in public planning. Common Ground, Melbourne

Healey P, Booher D, Torfing J, Sørensen E, Ng M, Peterson P, Albrechts L (2008) Civic engagement, spatial planning and democracy as a way of life civic engagement and the quality of urban places. Plan Theory Pract 9(3):379–414. https://doi.org/10.1080/14649350802277092

Hoch C (2011) What planners do. APA Press, Chicago

Holman P, Devane T, Cady S (eds) (2007) The change handbook. Berret-Koehler, San Francisco

Hurlimann A, March A (2012) The role of spatial planning in adapting to climate change. WIREs Clim Change 3:477–488. https://doi.org/10.1002/wcc.183

Information and Communications Technology Authority (ICTA) (2005) e-Government master plan in Mongolia. New York: United Nations Public Administration Network. Online resource, http://workspace.unpan.org/sites/internet/Documents/UNPAN042821.pdf. Last accessed 5/7/2017

Intergovernmental Panel on Climate Change (IPCC) (2001) Climate change 2001: the scientific basis. Cambridge University Press, Cambridge

Kerski J, Clark J (2012) The GIS guide to public domain data. ESRI Press, Redlands

Madera C (2010) Creating a culture of civic engagement. The transportation politic. Next City. Online resource, https://nextcity.org/daily/entry/creating-a-culture-of-civic-engagement. Last accessed 5/7/2017

Malik K, Wagle S (2002) Civic engagement and development. In: Fukuda-Parr S, Lopes C, Malik K (eds) Capacity for development. New solutions to old problems. United Nations Development Programme, New York, pp 85–100

Mandarano L, Meenar M, Steins C (2011) Building social capital in the digital age of civic engagement. J Plan Lit 25(2):123–135. https://doi.org/10.1177/0885412210394102

McArthur J (2017) Sidelining planners makes for poorer urban policy, and future generations will pay the price. The Conversation. Online resource, https://theconversation.com/sidelining-planners-makes-for-poorer-urban-policy-and-future-generations-will-pay-the-price-73186. Last accessed 5/7/2017. Also available from https://phys.org/news/2017-03-sidelining-planners-poorer-urban-policy.html

NORC (2016) Investigating how Americans navigate the modern information environment. NORC at the University of Chicago, Chicago

O*NET (2017) Quick search for urban planning. U.S. Department of Labor, Employment & Training Administration, Online resource, https://www.onetonline.org/find/quick?s=urban+planning. Last accessed 5/7/2017

Otto-Zimmermann K (2011) Embarking on global environmental governance. International Council for Local Environmental Initiatives (ICLEI), Bonn

Ramasubramanian L et.al (2016) Neighborhood and community perspectives of resilience in the Jamaica Bay watershed. In: Sanderson E, Solecki W, Waldman J, Parris A (eds) Prospects for resilience: insights from New York City's Jamaica Bay. Island Press, Washington, DC, pp 117–137

Rotondo F (2012) The U-City paradigm: opportunities and risks for E-democracy in collaborative planning. Fut Intern 4:563–574. https://doi.org/10.3390/fi4020563

Sarkissian W, Hurford D, Wenman C (2010) Creative community planning. Earthscan, Washington, DC

Showstack R (2014) Floods, climate change, and urban resilience: one policy Maker's perspective. Eos Trans AGU 95(24):201–203

Stern N, Taylor C (2007) Climate change: risk, ethics, and the stern review. Science 317(5835):203–204. https://doi.org/10.1126/science.1142920

Wilby R, Wigley T (1997) Downscaling general circulation model output: a review of methods and limitations. Prog Phys Geogr 21(4):530–548. https://doi.org/10.1177/030913339702100403

Wildavsky A (1973) If planning is everything, maybe it's nothing. Policy Sci 4(2):127–153

Chapter 2
Planning Challenges and the Challenges of Planning

2.1 Introduction

This chapter begins to articulate some answers to the question that is foremost in the minds of planning students and new practitioners: *Why is planning such a hot button issue?* For those of us who hold a worldview that is sympathetic to the benefits and advantages of planning in our day-to-day lives (and that probably includes you, because you picked up this book), it is often hard to understand why anyone or any group would think unfavorably about planning. This schism exists because planning for ourselves and our families is quite different from planning in the public realm. Planning with and for the public requires people with very different world views to work together to make decisions about the future. In this chapter, we look at the multifaceted nature of planning in the public realm, the substantive societal challenges that planners are confronted with in making specific decisions, as well as the challenges associated with the process of planning, affecting instrumental actions.

2.1.1 Making the Simple Complex

In the abstract, it is quite easy to say that we all, as human beings, want the same things – good jobs, clean water, safe streets, and so on. Politicians of all stripes are adept at speaking to these universal yearnings. Yet, public planners concern themselves about moving away from the abstract language of political slogans to address tangible concerns. Thus, a planner might ask, "how does one define a good job?" Surprisingly, there is more than one way to answer this question. One could argue that good jobs that those that pay a living wage, while another may propose that a good job is ideally one that provides a living wage and also includes guaranteed health care and other societal benefits. Others may add additional dimensions

L. Ramasubramanian, J. Albrecht, *Essential Methods for Planning Practitioners*, The Urban Book Series, https://doi.org/10.1007/978-3-319-68041-5_2

stating that a good job is one that also provides flex time and respects diversity in the workplace. The list can go on.

Undoubtedly, there are basic needs – many, especially those without work, would claim that having a job without health benefits is better than not having a job at all. So, for the sake of continuing this discussion and elucidating this argument, let's simply agree that a good job is one that provides a safe workplace, offers a livable wage, and provides some guarantees like paid sick leave. Having settled on the definition of a good job, a planner may ask another question; "how many of these "good jobs" are needed?" They may follow up with other, more complex questions, including trying to understand if the educational and skill levels of workers match the needs and expectations of prospective employers in a specific geographic area. They may try to assess if there is buildable land available nearby to situate a factory or an office building and whether there is supporting infrastructure to accommodate workers' families, schools for the workers' children, and hospitals to care for the workers and their families when they are ill. Thus, what began as a simple question becomes more complex and multilayered as planners get involved in the conversations. Rightly so. This is because of considerations of cost, specifically, the cost to the public purse. The public is supportive of almost all initiatives until they start to consider the thorny question, "How much is this all going to cost us?" We will return to the issue of cost and expenditure in a moment, but first, let's spend a little time thinking about how planners answer their detailed questions they pose to the public.

2.1.2 Data, Historical Trends, and Best Practices

Unlike conventional scientists and researchers, planners, whether they work for the government, consulting companies, or nonprofits, are always working against externally imposed deadlines. The cadence of decision-making is unpredictable, and often, there is an illusion that every decision is time-sensitive and urgent. Eventually, planners learn about project management, time budgeting, and prioritization, topics discussed in Chapter 7, but it's safe to say that time is always in short supply. In this situation, planners rely on data (about places or peoples), historical trends, and/or best practices (planning interventions that have worked effectively in similar social or geographical contexts) to arrive at some recommended action steps. We argue that practitioners should understand the strengths and limits of planning (from the past) so that they can plan more effectively for the future. Therefore, we briefly trace the planning shifts that have occurred in the past few decades as well as those demographic and societal challenges that are anticipated in the next two to three decades to provide practitioners with specific strategies to navigate the contemporary planning landscape. Section 2.2 focuses on some of these broad global historical and social trends. These trends will manifest differently across different geographical regions, states, communities, and neighborhoods. We anticipate that thoughtful planners will carefully link the trends we describe in Chapter 2 to the particularities of the situational contexts within which they work. To better understand our approach, it may be useful to spend some time reviewing the material we have provided about our two case study neighborhoods in New York, discussed in Chapter 3, which discuss the role of data and best practices in their situational context.

2.2 Planning Through the Ages

Planning and the development of cities/regions are interlinked; planning became necessary about the time when humans started to create permanent settlements (Mumford 1961). These settlement patterns were defined by many variables, such as the physical geography of the area, its climatic conditions, the materials that were available for construction, the origins of the people, and perhaps more importantly, the overarching purpose or need that guided the creation of a settlement. Different types of settlement patterns emerged and evolved over centuries. Even a cursory survey of human settlements worldwide will identify distinct typologies such as religious centers and temple towns, centers of power or capital cities, fortified cities, military encampments, cities for pleasure or recreation, and so on (Kostoff 1991). Early settlement patterns were usually self-contained and compact. There are many international examples of unique building typologies in these settlements (Cole 2002).

Most cultures/faiths used formal and informal and sometimes "religious" rules to organize their villages, towns, and cities, from the early civilizations of Mohenjo-Daro and Harappa (Lynch 1981). Typically, these settlements emphasized socio-spatial hierarchies and relationships. Of relevance to the today's planning practitioner is to note that in each of these settlement patterns, architects and planners used the technologies of their time to create "aesthetically pleasing" spatial arrangements (layouts) that fulfilled a variety of practical needs (Rapoport 1969). However, the variability of these spatial arrangements suggests that ideologies related to culture, power, and faith also shaped the physical form of these settlements. The manipulation of individual buildings through their size, scale, proportion, and massing, as well as the use of materials, colors, and textures, provided a nonverbal communication of the dominant social order (Brown 1942; Kostoff 2010). The relationship of buildings to one another, along with the relationship of buildings to pathways and byways that connect them, communicate a great deal of information about the dominant social structure and functions of a society at any given time during its development. Additionally, there is a recursive relationship that settlement patterns, once established, continue to perpetuate thereby reifying prevalent patterns of social hierarchies and interactions. Changes in the morphology of human settlements can also communicate information about how/where the town expanded and perhaps yield clues to why the expansion followed specific pathways (Knox 2014). In cities across the world, architects, urban designers, and planners, either working alone or with the support and encouragement of civic/religious leaders, have articulated a social order that was clearly legible in the spatial order (organization) of the settlement (Bacon 1976; Kostoff 1991).

While there are some similarities, there are also sharp differences in settlement patterns in different countries. For the most part, this chapter and the book focus on the development of settlements in the United States. However, it is important to note that these US settlement forms discussed in the next few sections have been "exported" to many other countries. Non-Western social and cultural milieus in some of the most populated areas of the world like India and China continue to create "globalized" planning visions in the development of their own cities using planning ideas developed in the west (Planning Commission, Government of India 2013; Wu 2015).

2.2.1 Twentieth-Century Settlement Patterns in the United States

In the United States, as in much of Western Europe, early twentieth-century cities responded to the challenges of early industrialization. Early settlements included mercantile and port cities, industrial company towns, state capitals, and new frontier towns (Hall 1988). While early industrialization created many challenges to human health and well-being, the development of new materials and construction techniques established new building forms and styles that we still recognize today (Miller 1996; Larson 2003). Settlements created for one reason or another evolved, expanding and adapting to demographic growth and change (Davis 1965). Transportation and available transportation technologies circumscribed and limited the expansion and growth of early settlements, benefiting the growth of some cities while inhibiting others (e.g., Gurda 1999). However, the advent of the automobile and the growth of robust road networks created new patterns of development soon after World War II (Taylor 1998).

The expansion of compact urban settlements, spurred by investments in new transportation infrastructure, facilitated and expanded suburbanization. Much has been written about suburbanization in the United States and its consequent impacts on urban development (see, e.g., Jackson 1985; Hayden 2003). Suburban development was characterized by neighborhoods comprising of spacious single-family homes that included private open spaces and created a sense of security through the layout and arrangement of individual homes along streets (cul-de-sacs). Private automobile ownership and transportation connectivity also determined the individual house form (prominent garage/driveway). The hetero-normative household dominated the social order. The prevailing societal attitudes assumed that heads of households (usually men) commuted between work (in the city) and home (in the suburb) (Gans 1969; Stimpson et al. 1981). Larger structural forces also shaped the vision of a new social order during this time.

The planning field has always been a multidisciplinary community that includes architects, landscape architects, engineers, and social scientists. Planning practitioners organized themselves as early as 1917 as the American City Planning Institute. The American Institute of Planners (the precursor to the present-day American Planning Association) was established in 1939.[1] Dominant planning trends have

---

[1] American Planning Association: History. Available at: https://www.planning.org/apaataglance/history.htm. Retrieved March 1, 2017.

always been shaped by technological and political developments. Different ideas about planning have been put forward over the last 100 years, reflecting the diversity of thought and intellect of the times. These divergent philosophies continue to evolve. There can be no objective retracing of the evolution of American planning practice. Suffice it to say that a wide range of political and social ideologies have always been embedded in planning practices, then, and now.

One of the recurrent themes that dominate international planning practice is the desire to create a settlement from a clean slate – a tabula rasa. In so far, that it is an impossible aspiration, it is also a planning ideal.[2] Historically, architects and planners have always proposed "ideal" settlements or "planned communities."[3] Examples abound of such visionary initiatives, including a case study that we include in Chapter 3 – Roosevelt Island.

2.2.2 *Changing Morphologies, Urban Sprawl, and New Urbanism*

The suburban form that was established in the 1950s and 1960s has continued to grow and expand, and, at least, two different orders of suburban development can be distinguished; an earlier modestly scaled type of suburban development has given way to newer developments that are oversized with an added emphasis on exclusivity, achieved through the creation of gated communities (Blakeley and Snyder 1997; Low 2003). The 1980s also saw the emergence of *Edge Cities* (Garreau 1991) that helped reduce the prominence of a traditional central city. Edge cities provide city-like amenities (offices, retail environments, hotels), direct connections to suburban residential zones, as well as access to a major airport, allowing residents to avoid the central city altogether. Edge cities are known to have a distinct (named) identity that identifies it as a business center.

Despite planners' affection and continued interest in living and working in the traditional city, we encourage young planning professionals to critically examine suburbanization trends in the United States as well as other parts of the world.[4] American suburbs deserve to be examined and studied with the same enthusiasm that is typically reserved for cities. About half of all Americans live in a suburban community, and we can speculate that they live there by choice (albeit balancing trade-offs associated with housing prices, commuting costs, and other variables).[5]

---

[2] *Beijing's new annex: A plan to build a city from scratch that will dwarf New York, The Economist*, April 6, 2017, Available at: http://www.economist.com/news/china/21720318-will-xi-jinpings-dream-come-true-plan-build-city-scratch-will-dwarf-new-york. Retrieved April 6, 2017.

[3] Don, K. 2010. *Frank Lloyd Wright's Utopian Dystopia*, April 8, 2010, Available at: https://nextcity.org/daily/entry/frank-lloyd-wrights-utopian-dystopia. Retrieved Feb 2, 2017.

[4] *The Economist Essay: A Planet of Sub*urbs Not Dated. Available at: http://www.economist.com/suburbs. Retrieved March 2, 2017.

[5] US Census Bureau, 2002. Demographic Trends in the Twentieth Century, Census 2000 Special

Yet, American suburbs vary greatly in their economic viability – pockets of poverty, deteriorating housing stock, and high incidences of crime are the hallmarks of some suburban communities. Progressive planners should pay attention to the challenges facing the ring of older suburbs that are in the immediate periphery of older cities in the Northeast and the Midwest.[6] Immediately following the financial crisis of 2008, some new planned suburban communities were transformed into a desolate landscape of suburban blight[7] almost overnight. By 2017 that trend seems to be reversing itself.[8]

A countervailing trend of planned urban development (that attempts to bridge the city-suburb divide) is the creation of neighborhoods based on *new urbanist* principles – i.e., principles of traditional neighborhood design, a settlement form that was prevalent before the rise of the automobile (Duany et al. 2000). The development of New Urbanism as a planning framework has forced American urban planning practitioners to consider the physical form of buildings, turning the old dictum – *form follows function*[9] – on its head. New urbanist principles include a vision of walkable, pedestrian-friendly neighborhoods that provide many opportunities for social interaction (Duany and Plater-Zyberk 2003). New Urbanism also emphasizes the creation of a legible neighborhood with a clear "center and an edge" that is no more than a 10-min walk away. Within these legible (easy to navigate) neighborhoods, a variety of residential and compatible nonresidential uses are encouraged. New urbanists pay attention to the placement of buildings and the relationships of buildings to the street, scaling buildings to human scale and minimizing the importance given to the private automobile.

Some academic scholars observe that New Urbanism is neither new nor urban,[10] reminding us that many new urbanist principles harken back to the settlement patterns of small villages and country towns in a pre-automobile era. Others point out that most tangible applications of the concept can be observed in planned residential settlements. Still others label it social engineering. New urbanist theories and methods received political support when the US Department of Housing and Urban Development under the leadership of Henry Cisneros accepted new urbanist ideals

---

Reports. Available at: https://www.census.gov/history/pdf/1970suburbs.pdf. Retrieved March 2, 2017.

[6] Puentes, R. and D. Warren, 2006. *One-Fifth of America: A Comprehensive Guide to America's First Suburbs*. Washington, D.C.: Brookings Institution Survey Series Available at: https://www.brookings.edu/wp-content/uploads/2016/06/20060215_FirstSuburbs.pdf. Retrieved March 2, 2017.

[7] Kilston, L. 2013. Economic collapse seen through aerial photos of abandoned mansions. Available at: https://www.wired.com/2013/09/michael-light-aerial-photos/. Retrieved March 2, 2017.

[8] Cox, W. 2017. Flight from urban cores accelerates: 2016 census metropolitan area estimates. Available at: http://www.newgeography.com/content/005570-flight-urban-cores-accelerates-2016-census-metropolitan-area-estimates. Retrieved on March 26, 2017.

[9] Attributed to American architect, Louis Sullivan.

[10] Dunham-Jones, E. 1998. Academics take a hard look at the New Urbanism, Public Square, November 1, 1998. Available at: https://www.cnu.org/publicsquare/academics-take-hard-look-new-urbanism. Retrieved March 1, 2017.

in the planning and revitalization of inner-city communities in the early 1990s to address some of the endemic problems associated with public housing developments (Cisneros and Engdahl 2009). Despite the fact that the theory has some detractors, it is useful to note that no "new" or "radical" theories or principles of physical planning have been put forward to pragmatically guide urban settlement development since the emergence and growth of New Urbanism.

2.3 Planning Challenges

When we plan for a short time frame, there is a high level of predictability and reliability in the success of the plan. However, there is no real benefit in planning for the next hour or next day in a very high degree of detail except for certain specialized situations such as disaster response and recovery. Yet, when we increase the time horizon for planning to attempt to anticipate problems that may occur in the next 50 or 100 years, we must cope with a great deal of ambiguity. However, practicing planners must recognize that the following three major future trends will impact almost every aspect of their work. They include urbanization, demography, and climate change. Each element creates a transformation – changes in the physical landscape, changes in the population, and changes in the environment. Collectively, these changes create new realities on the ground that should be included in the planning and management of any program, project, or policy.

2.3.1 Urbanization

Urbanization, in planning terms, reflects the reallocation of land to urban uses like housing, infrastructure, and transportation as opposed to non-urban uses such as farming or leaving land in its natural state. Urbanization results from population growth through natural cycles of reproduction and/or the migration of populations from rural to urban areas. Cities are growing worldwide, and new megacities (cities with ten million or more in population) have increased from 10 cities in 1990 to 28 cities in 2014.[11] Growth is not restricted to megacities – smaller towns and cities are also growing and urbanizing steadily.[12] The United States is likely to follow global urbanization patterns removing conventionally understood distinctions between cities and suburbs.[13]

[11] World Urbanization Prospects 2014 Revision, United Nations. Available at: https://esa.un.org/unpd/wup/publications/files/wup2014-highlights.Pdf. Retrieved on February 1, 2017.

[12] The World's Cities in 2016. Data Booklet, United Nations. Available at: http://www.un.org/en/development/desa/population/publications/pdf/urbanization/the_worlds_cities_in_2016_data_booklet.pdf. Retrieved on February 1, 2017.

[13] Berube, A. 2011. The State of Metropolitan America: Suburbs and the 2010 Census. Available at:

2.3.2 Demography

The American population is growing steadily, and the country is projected to have over 400 million people by 2051.[14] While the birth rate has been declining since the 1950s, Americans are living longer. At the same time, it appears that many young Americans are delaying marriage and parenthood.[15] These trends are expected to continue. America is also becoming a more diverse country, with millennials, young adults between the ages of 18–34, accounting for much of this diversity. By 2050, America is likely to become a "majority-minority" nation.[16]

2.3.3 Climate

Most scientists agree that the earth's climate is changing rapidly[17,18] (see Fig. 2.1). Naturally occurring fluctuations have accelerated and increased by orders of magnitude because of human interventions such as population growth, urban development, and environmental degradation. The impacts of human-induced climate change are likely to be highly variable and have localized and disparate consequences; for example, coastal areas are likely to be severely affected by flooding because of rising sea levels (Rosenzweig and Solecki 2015).

2.4 Impacts and Consequences

As a planning professional, it is important to consider aforementioned global trends and understand how they are likely to impact planning initiatives or actions within the geographical region where you work. Some general impacts and consequences

---

https://www.brookings.edu/on-the-record/the-state-of-metropolitan-america-suburbs-and-the-2010-census/. Retrieved on June 1, 2016.

[14] Colby, S and J. M. Ortman, 2015. Projections of the Size and Composition of the U.S > Population: 2014 to 2060. Available at: https://www.census.gov/content/dam/Census/library/publications/2015/demo/p25-1143.pdf.Retrieved on June 1, 2016.

[15] Vespa, J. 2017. The Changing Economics and Demographics of Young Adulthood: 1975-2016. Available at: https://www.census.gov/content/dam/Census/library/publications/2017/demo/p20-579.pdf. Retrieved on April 15, 2017.

[16] Colby, S and J. M. Ortman, 2015. Projections of the Size and Composition of the U.S > Population: 2014 to 2060. Available at: https://www.census.gov/content/dam/Census/library/publications/2015/demo/p25-1143.pdf.Retrieved on June 1, 2016.

[17] Global Climate Change: Vital Signs of the Planet. Available at: https://climate.nasa.gov/evidence/. Retrieved on May 1, 2017.

[18] IPCC, 2014: Climate Change 2014: Synthesis Report. Contribution of Working Groups I, II and III to the Fifth Assessment Report of the Intergovernmental Panel on Climate Change [Core Writing Team, R.K. Pachauri and L.A. Meyer (eds.)]. IPCC, Geneva, Switzerland, 151 pp. http://www.ipcc.ch/report/ar5/syr/. Retrieved on January 2, 2016.

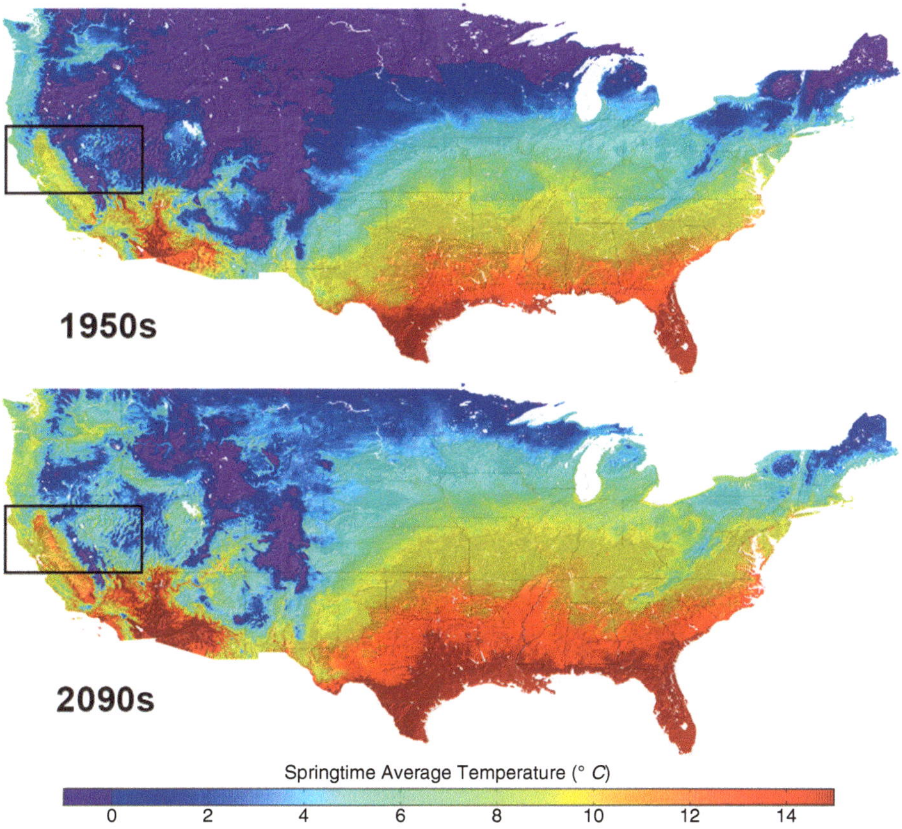

Fig. 2.1 Springtime temperatures are predicted to increase significantly (Image credit: NASA)

of urban growth, demographic change, and climate change are discussed below and depicted in Fig. 2.2 to help jump-start your reflections. These considerations should influence the specifics of the design, planning, and management of programs, projects, or policies.

2.4.1 *Combat and Manage Sprawl*

One of the main planning challenges of urban development in the United States is sprawl (Gillham 2002). Sprawl, concisely defined, is the rapid growth of low-rise, single-family housing developments. Conservatives and liberals debate the causes, extent, and the problems of sprawl (Bruegmann 2005; Squires 2002).

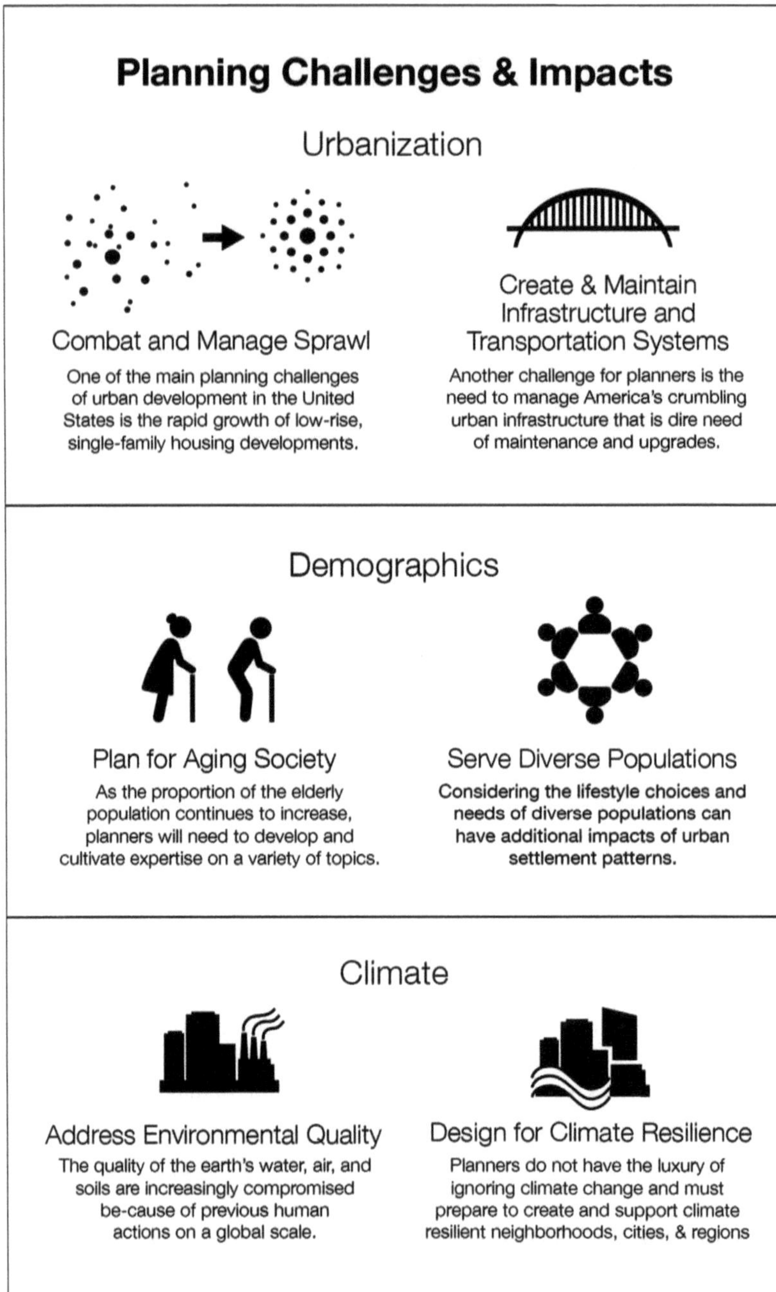

Fig. 2.2 Planning challenges and impacts

Even those who want to limit and curb sprawl are divided about to achieve their objectives.[19] Policy and design solutions have been put forward, but the skepticism of the public toward growth management through regulations prevents these methods and approaches from being implemented on a grand scale (Downs 2005; Knapp and Talen 2005). *Smart growth principles*, conceptualized as an antidote to sprawl, encourage the provision of better public transportation options, new zoning regulations to increase density where possible, as well as investments in mixed-use town centers as a part of residential developments. Other smart growth principles create disincentives to sprawl by transferring some of the costs of infrastructure development (road networks, water and sewer connections, etc.) to the end users and by imposing more stringent development protocols (Ingram and Hong 2009).

2.4.2 Create and Maintain Infrastructure and Transportation Systems

Another challenge for planners is the need to manage America's crumbling *urban infrastructure* that is dire need of maintenance and upgrades.[20] Infrastructure is the generic term given to road and rail networks, bridges, drinking water and sewer systems, flood management systems like dams and levees, as well as facilities that are necessary for a vibrant and thriving economy such as aviation terminals and ports. Public facilities like schools, hospitals, parks, and playgrounds are also part of the urban infrastructure. As public infrastructure ages, various systems break down or are closed voluntarily to address safety concerns. In addition to the serious cost of lost human lives, productivity declines. Economic growth can be stalled because of the infrastructure crisis. Some argue that the country's older cities and neighborhoods are worst affected[21] although the challenges faced by smaller communities do not gain visibility until the situation reaches crisis proportions.[22] The management of urban infrastructure is complicated by the fact that much of it is invisible (consider water, sewer, and utility lines) and/or falls under different jurisdictional authorities (consider rail networks).

---

[19] Gordon, P & H.W. Richardson,1998. Prove It: The Costs and Benefits of Sprawl. Brookings Review, Fall 1998, Available at: https://www.brookings.edu/wp-content/uploads/2016/06/gordon2.pdf. Retrieved on June 12, 2016.

[20] 2017 Infrastructure Report Card, American Society of Civil Engineers, Available at: http://www.infrastructurereportcard.org/. Retrieved April 2, 2017.

[21] Fishbein, R. 2014. NYC Infrastructure is crumbling, March 12th, 2014, Gothamist Available at: http://gothamist.com/2014/03/12/nyc_infrastructure_rip.php. Retrieved September 1, 2016.

[22] Dixon, J. How Flint's water crisis unfolded. Detroit Free Press. Available at: http://www.freep.com/pages/interactives/flint-water-crisis-timeline/. Retrieved on April 30, 2017.

2.4.3 Plan for an Aging Society

One of the biggest impacts of the demographic shifts sweeping the country relates to the "graying" of America with the population of people 65 years and over projected to become about 98 million by the year 2060.[23] As the proportion of the elderly population continues to increase, planners will need to develop and cultivate expertise on a variety of topics[24] related to aging in place. Some of these topics include housing for aging populations, including the retrofitting and adaption of existing housing stock to accommodate elderly homeowners and home buyers;[25] transportation services, including the management of private transportation services for individuals and small groups, particularly in low-density neighborhoods (Wachs 1979); and the development and maintenance of residential health-care facilities and ancillary nonresidential options such as elder day-care[26] (see also Fig. 2.3).

2.4.4 Serve Diverse Populations

Diversity often references racial and ethnic differences, but it should consider much more, including age, gender, sexual orientation, marital status, and different physical and mental abilities. For example, considering the lifestyle choices and needs of young adults can have additional impacts of urban settlement patterns; young adults seem to prefer compact living arrangements in urban areas where they continue to seek out a range of entertainment, retail, and dining options.[27] They are more likely to put off making major purchases like a car or their first home, thereby forcing planners to examine the benefits and consequences of an expanding rental housing market and expanding public transportation infrastructure. Planners must consider the complexity of planning and designing for diversity to ensure that our physical fabric is accessible and inclusive.[28]

---

[23] United States Department of Health and Human Services, 2016. A Profile of Older Americans: 2016. Washington, D.C: Administration for Community Living Available at: https://aoa.acl.gov/Aging_Statistics/Profile/index.aspx). Retrieved April 15, 2017.

[24] American Planning Association: Aging in Community Policy Guide. Available at: https://www.planning.org/policy/guides/adopted/agingincommunity.htm. Retrieved on March 12, 2017.

[25] Aging in Place: facilitating choice and independence https://www.huduser.gov/portal/periodicals/em/fall13/highlight1.html.

[26] Ball, M.S. not dated. Aging in Place: A Toolkit for Local Governments. Available at: http://www.aarp.org/content/dam/aarp/livable-communities/plan/planning/aging-in-place-a-toolkit-for-local-governments-aarp.pdf. Retrieved on May 10, 2016.

[27] Vespa, J. 2017. The Changing Economics and Demographics of Young Adulthood: 1975-2016. Available at: https://www.census.gov/content/dam/Census/library/publications/2017/demo/p20-579.pdf. Retrieved on April 15, 2017.

[28] Winograd, M & M. Hais. 2014. Howe Millennials could upend Wall Street and corporate America. May 28th, 2014. Available at: https://www.brookings.edu/research/how-millennials-could-upend-wall-street-and-corporate-america/. Retrieved on September 30, 2016.

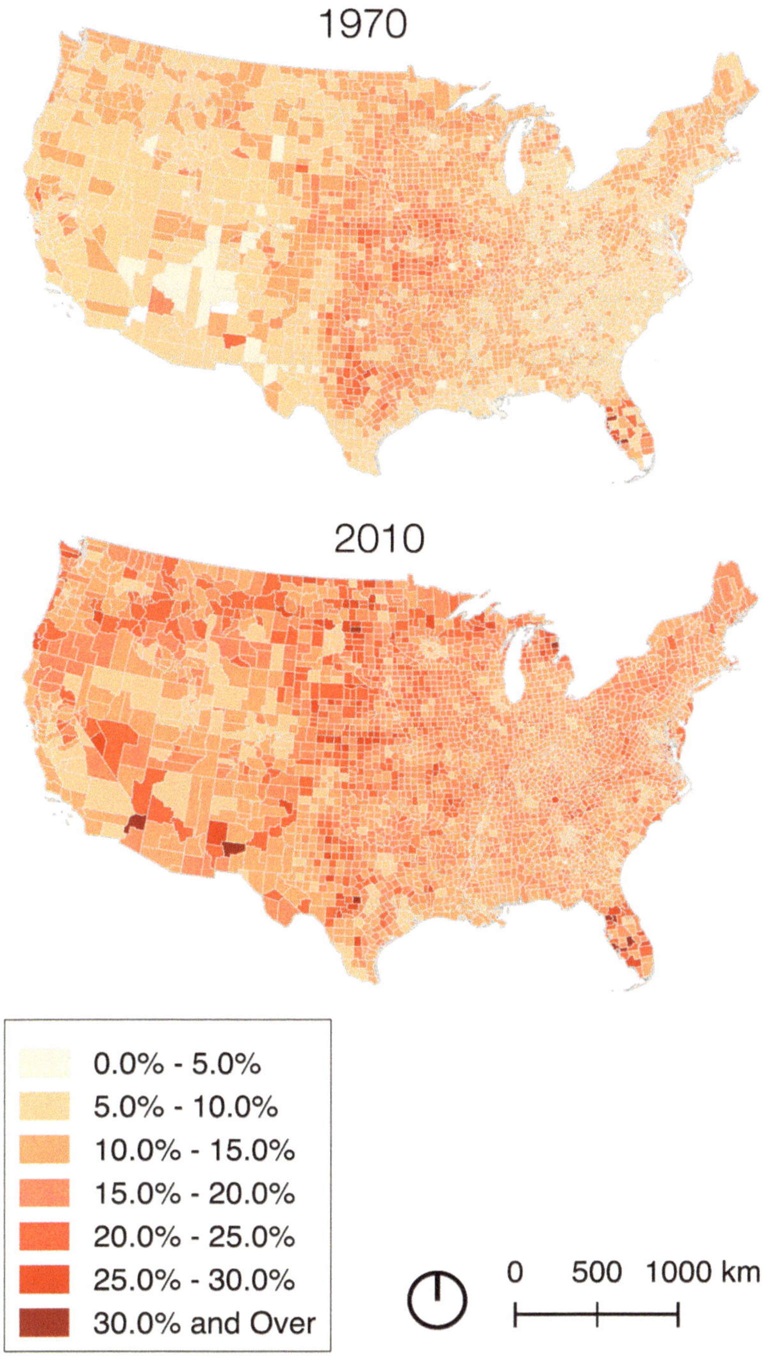

Fig. 2.3 An aging society. The percentages refer to percent of county population age 65 years or older

2.4.5 Address Environmental Quality

According to many governmental and non-governmental agencies around the world, the quality of the Earth's water, air, and soils is increasingly compromised because of previous human actions on a global scale. For example, the air quality of many world cities is compromised because of the presence of particulate matter, ozone, and other global greenhouse gases such as methane, nitrous oxide, and chlorofluorocarbons.[29] Water quality of lakes, rivers, and streams is affected by agricultural runoff that is loaded with chemical fertilizers and pesticides. Massive storms and heavy rainfall result in sewer overflows that contaminate the environment. Soils can also be contaminated because of different types of extractive activities such as mining or drilling that generate polluting by-products. Illegal dumping of wastes can further contaminate the soil. Pollutants can work their way through the soil and contaminate groundwater sources as well. Practitioners should prepare themselves by learning different strategies of ensuring environmental quality, particularly to identify strategies to increase the public trust that the environment is safe for human habitation.[30]

2.4.6 Design for Climate Resilience

Planners do not have the luxury of ignoring climate change and must prepare to develop and support climate resilient neighborhoods, cities, and regions (Bicknell, et.al 2009). The impacts of climate change are highly localized. Disparate impacts create "winners" and "losers" in the short term. Developing plans for climate resiliency requires planners to understand and interpret the science behind climate data to advocate for practical solutions that can protect private housing stock and public facilities, including transportation infrastructure.[31] These are not limited to physical interventions – planners must also consider policies such as overhauling city and state zoning codes. Planners must also develop response strategies to prepare and manage the immediate and long-term impacts of extreme weather events that are likely to occur more frequently in the next 50–100 years.[32]

---

[29] IPCC, 2014: Climate Change 2014: Synthesis Report. Contribution of Working Groups I, II and III to the Fifth Assessment Report of the Intergovernmental Panel on Climate Change [Core Writing Team, R.K. Pachauri and L.A. Meyer (eds.)]. IPCC, Geneva, Switzerland, 151 pp. http://www.ipcc.ch/report/ar5/syr/. Retrieved on January 2, 2016.

[30] Flint Water Crisis Fast Facts. Available athttp://www.cnn.com/2016/03/04/us/flint-water-crisis-fast-facts/. Retrieved April 30, 2017.

[31] Jarvis, B. 2017. When Rising Seas Transform Risk into Certainty, The New York Times Magazine, April 18, 2017. Available at: https://www.nytimes.com/2017/04/18/magazine/when-rising-seas-transform-risk-into-certainty.html?hp&action=click&pgtype=Homepage&clickSource=story-heading&module=photo-spot-region&region=top-news&WT.nav=top-news&_r=1. Retrieved on April 20, 2017.

[32] Rice, A. 2016. This is New York in the not-so-distant future, September 5th, 2016, The New York Magazine, Available at: http://nymag.com/daily/intelligencer/2016/09/new-york-future-flooding-climate-change.html. Retrieved on January 30, 2017.

2.5 Challenges of Planning

Section 2.3 discussed three global trends, urbanization, demography, and climate change, and Sect. 2.4 presented six planning challenges that can arise as a result. In most democratic societies, planning is a social and political activity that engages a variety of stakeholder groups. When planners engage the public to develop plans, proposals, and policies, they encounter a different set of challenges. We refer to them in this book as the challenges of planning,[33] in other words, process-oriented challenges that are different from the planning challenges described in Sect. 2.4.

2.5.1 The Death of Expertise[34]

Planners are professionals. Most planners would claim that they are "experts" in that they are armed with technocratic knowledge about how to create and manage different aspects of our society. Planning education and planning practice are specialized in several ways. For example, a planner may develop expertise in housing, economic development, or transportation. They could further be an expert in housing design, transportation modelling, or the use of tax policy to spur economic development. While individual expertise is cultivated over time, it is usually informed through rigorous and systematic study of the subject matter and informed with on-the-job experience that is gathered through fieldwork placements or internships. Yet, our society has changed in dramatic ways – the democratization of information, spurred by digital technologies and the Internet, have allowed everyday people to gain expertise and knowledge quickly, often without following the traditional pathways of a rigorous planning education. Planners engaging with the public often find their own expertise challenged by laypeople (and politicians) who want quick fixes to pressing urban problems. Planners often advocate reasoned and thoughtful responses to situations and ask a lot of questions (see Sect. 2.1 of this chapter). They often encounter resistance from the public who may view delays as stalling or a desire to avoid decision-making. This is a complex challenge for planners who want to be participatory and be engaged with their publics and yet carve a role that they can play. Are planners merely highly paid facilitators of public dialogue? What exactly is their contribution in shaping the physical and social fabric of our cities and towns? This is a question that we will return to, in subsequent chapters.

---

[33] These ideas were first publicly presented by Dr. Laxmi Ramasubramanian at a seminar at Virigina Tech in February 2017.

[34] We encountered this idea explored in a different context in Tom Nichols blog post in The Federalist. Available at http://thefederalist.com/2014/01/17/the-death-of-expertise/. Retrieved on June 25, 2016.

2.5.2 Planning with Diverse Populations

Planning has always been a blend of the technical and the political. Planning visions are crafted by the political power elite from time immemorial and continues to the present day. Planning in many societies is managed in a top-down hierarchical fashion, and the citizens and the public acquiesce or accommodate these imposed views. Planning implementation – the aspect of planning that is most visible in the public realm – is streamlined. Words and phrases like "efficient," "quick," and "completed on time and under budget" signify an orderly, rational, technocratic process that serves all people equally. This vision of planning is only an illusion, particularly in democratic and open societies. Planning is accountable to, and serves many interest groups who compete and jockey for power and influence over the decision-making process, and the final outcomes. Clear mandates seldom emerge and planners negotiate between and among competing interest to advance projects or policies. In their day-to-day work, planners often develop proposals or recommend outcomes that can satisfy the needs of a considerable majority of the residents in a neighborhood or city. They strive to build compromise to encourage public engagement, commitment, and support of their initiatives. These efforts are more successful when there is broad agreement about what is practical and feasible. Yet, our society has become hyper-diverse, not simply in racial or ethnic terms but in every aspect of life. In fact, being different is valued more than conformity. This is not inherently a problem, but it makes consensus building harder. In a community that includes homeowners, renters, business people, young and old, and rich and poor, it is harder to arrive at an agreement about what is the most appropriate way forward. This is a good problem to have, for planners – but still a problem that must be confronted head on. Practicing planners that work in neighborhoods or cities with large immigrant populations should anticipate that they will work on a range of issues related to social services provisions, education, housing, as well as economic development. For instance, ensuring that information is available in languages other than English, many eventually become necessary in many communities.

2.5.3 The Future of Participation

For the last 50 years, planners have strenuously supported the ideal of planning with, rather than for, the public. An important ingredient in planning processes is the attention given to public participation. Participation is embedded into the institutional fabric of planning initiated by the federal government, and to the extent that federal dollars' flow into states and local communities, stringent participation guidelines often follow. In addition, most states and localities encourage some form of democratic participation through information sharing or community consultations. Academic planners have debated and reflected on the nature of public participation in public planning for many decades. A plethora of best practices to facilitate and manage participation has been assembled over this time. Yet, many would argue that public participation processes have not always improved the quality of the outcomes, contribute to project delays, and are often used as political theater. Planners of the future should engage critically with the question: What forms of public participation are necessary to address the challenges of the future? What forms of community engagement are necessary to invite and engage individuals and groups who are perpetually left out of decision-making loops? How can a broadly consultative and collaborative planning process become sustainable financially and avoid burn out among planners and the public alike? Some of these questions are discussed at length in Chapter 6.

2.6 Technologies and Planning Practices

The development of planning practices in the United States after World War II was strongly influenced by a sense of optimism and grounded in the ideals of technology-driven progress. Nowhere have these impacts been more visible and more contentious than in land use and transportation planning (Plummer undated, modified 2007), although it has affected many other sectors including housing, economic development, and public health. The use of quantitative data and mathematical models to explain and predict human behavior and the use of statistically significant analyses have been an integral part of American and Western planning since the end of World War II (Barnes 2003). The advent of the computers sped up this trend.

2.6.1 Planning as a Science

During the 1950s and 1960s, federal, state, and local government agencies emphasized large-scale, comprehensive planning projects. In order to plan and manage for rapid urban growth, planners amassed and analyzed a large volume of data about historical and current land use and transportation trends which they then used to forecast future patterns of growth. For example, travel demand forecasting, developed by the Chicago Area Transportation Study, used systematic procedures to compute trip generation, the modal split, trip distribution, and mode assignment (Black 1990). The planning goals identified by the agency, in consultation with community leaders, emphasized speed and efficiency. Then computerized and automated methods optimized route selection between destinations to identify the shortest travel paths with minimum impedances. The computationally intensive approaches served their purpose in some ways but neglected to consider the quality of the travel experience, the impacts on neighborhoods and communities that were not along the connected nodes of a regional network of expansion and development, and the concerns about sprawl that were discussed earlier.

Computational advancements and modeling techniques made it possible for practitioners to think and plan at the scale of the region. Planners, at the time, also made assumptions about societal and economic trends which influenced the identification of planning goals. For example, the locus of economic opportunity was situated as a fixed zone in the central city; thus, an important planning goal was to reduce the commute time between the residential suburb and the central business district (CBD). The prevailing vision and value system – working men, stay-at-home women, a nuclear family as a unit of social interactions that occurred in the private sphere, workplaces situated in the central city, the utility, convenience, and status of owning a private automobile – contributed to the land-use patterns that emerged. These patterns were supported by financial incentives through mortgage programs provided by the government that encouraged homeownership as a pathway to opportunity, a continuing meme in American society that persists even today. While many of these projects and plans were initially seen as very successful, a more nuanced assessment is necessary. We propose that many of these comprehensive planning projects helped to create and support a new American middle class. However, these plans and policies also sowed the roots of income inequality and racial disparity that would explode in the 1960s across the United States.

The idea that planning was a science became gradually discredited by the late 1970s (Friedmann 1987; Taylor 1998). One can speculate that academics and practitioners began to recognize that plans they had developed using highly computationally intensive models and projections had helped create hardships and harm. Specifically, the development of highways and freeways (designed to speed up travel) and mega-development projects tore through urban neighborhoods where poor people and people of color lived. The argument that some people would inevitably have to make sacrifices for the greater benefits to the region failed to gain traction as the numbers of those displaced grew, and patterns of displacement began to become more visible. Across the country – Boston, New York, Chicago, and San Francisco, to mention a few cities – active resistance to expert-driven planning and/or top-down planning broke out (e.g., King 1981; Mollenkopf 1983).

2.6.2 Planning Support Systems

The development of personal computers and geographic information systems (GIS) re-energized spatial planning. Although tracing the history of GIS is beyond the scope of this chapter, it is important to note that parallel developments in fields allied with planning specifically, landscape architecture, and cartography helped to create principles, methods, and techniques for creating digital maps using locational (latitude, longitude) information during the 1960s and 1970s. As GIS technologies evolved, the United States Geological Survey (USGS) and the United States Census Bureau created first the standards of digital cartographic data and then used those standards to create digital geographic base maps.[35,36] During the 1980s, the process of analog to digital conversion of maps began, and maps were stored as digital data files. The creation of relational databases with location identifiers allowed different types of end users to use a locational reference such as a street address to connect different types of information that were available about a unique address regardless of which agency/group had been involved in collecting that information.

While paper maps have always been used for planning purposes, computerized mapping "disrupted" the status quo in planning agencies. There is a robust literature from planning practice that documents how different types of planning agencies adopted and adapted the newly emerging GIS technologies to support and advance their work (Huxhold 1991; Campbell and Masser 1995). City planners found that with the help of GIS, they could use the information that was generated by other city departments to improve efficiencies in routine tasks, make better management decisions, and create better policies. Many progressive outcomes resulted from the use of GIS in government, including a more equitable allocation of resources and services that were more appropriate to the needs of one particular community. GIS has also been used to identify instances of disproportionate burdens experienced by people of color because maps made using GIS could demonstrate that locally unwanted land uses (LULUs) had been placed in poor or minority neighborhoods by overlaying sociodemographic information with land use and facilities information (Ramasubramanian 2009).

Many nontechnical users are fascinated by the visual map displays made possible through GIS – while conventional paper maps are static, GIS software facilitates the creation of dynamic maps that allow for display, toggling on/off different types of information and features of a landscape, as well as showing changes that occur over time. It soon became apparent that planning departments and agencies that had access to GIS could make crisper and better-formed arguments to support their claims. As the "official" use of GIS expanded, non-governmental organizations and activists' groups began to take note (Ramasubramanian 2009).

2.6.3 *Participatory Planning Technologies*

A countermovement to "democratize" GIS began in the mid-1990s as the technology became popular, in part, through the efforts of a major software developer[37] who thoughtfully and consciously engaged universities, colleges, and schools. Partnerships with universities and community-based organizations allowed for a

---

[35] Dual Independent Map Coding Available at: https://www.census.gov/history/www/innovations/technology/dual_independent_map_encoding.html. Retrieved on April 3, 2017.

[36] TIGER = Topologically Integrated Geographic Encoding and Referencing Available at: https://www.census.gov/geo/maps-data/data/tiger.html. Retrieved on April 3, 2017.

[37] Jack Dangermond, UCGIS Fellow Citation. Available at: http://www.ucgis.org/jack-dangermond. Retrieved on April 9, 2017.

wide range of community-university partnership projects to use GIS tools and publicly available census data to ask more place-specific and issue-specific questions. In the 1990s, as the hardware and software became more affordable and user-friendly, the challenge that slowed everyone down was the lack of access to useful data. Undoubtedly the census provided a great deal of useful information, but the more interesting data sets that were useful to planners such as land-use information, property ownership information, and other business information were collected by different entities at different times and not generally available for use to the public.

Collecting and assembling data were a big stumbling block for geographic information systems to become community information systems. The push to create a National Spatial Data Infrastructure, along with a National Geospatial Data Clearinghouse, came as early as 1994 (Federal Geographic Data Committee 1994). The development and rapid growth of the World Wide Web in the mid-1990s added a new dimension to the adoption and use of GIS. As part of the development of the NSDI, government data and data collections held in public and university libraries and other locations became accessible through the creation of powerful map and data servers. In some instances, even if the actual data was not downloadable, non-technical users could search metadata (data about data) to identify useful and relevant information for their own specific needs. GIS adoption and use in government, business, nonprofits, and academia have continued to grow rapidly. We argue that a new wave of technical-rational planning was reestablished in the mid-1990s, although it was disconnected from a robust theoretical or ideological framing.

The Obama Administration's open data policies established in 2009 and further expanded in 2012 explicitly made open and machine-readable data the new default for government information.[38] The federal push to make data available and accessible as a new kind of infrastructure to the tech sector (application developers, civic hackers, and the like) has expanded to state and local governments.

Participatory GIS methodologies also benefited from the development of cloud-based computing services that moved the analysis away from desktops into virtual data servers and portals. Generally, tedious routines previously necessary to execute simple analyses have been replaced by a push-button interface design that makes the computation "invisible" focusing instead on the "visualization" of the results. For example, the advantages of Web interfaces that provide real-time traffic information allow users to make decisions and change their routes while in transit. The real-time information is generated in a variety of ways, most significantly through crowd-sourcing, where individual "smart" phones with location identifiers passively transmit data to a centralized server. An entire industry has emerged around "location-based" services that are available to end users for free or at very low cost.

[38] Obama Administration Executive Order – Making Open and Machine Readable the new default for government information, Available at: https://obamawhitehouse.archives.gov/the-press-office/2013/05/09/executive-order-making-open-and-machine-readable-new-default-government-. Retrieved on April 10, 2017.

2.6.4 Big Data, Social Media, and Planning Apps

"Big Data" is a phrase that speaks to the large volume and variety of data that is available in real-time or nearly close to real-time. Big data is the by-product of the popularity of powerful mobile phones throughout the world, supported by powerful data servers that can store and disseminate large volumes of information to individual users (Kitchen and McArdle 2016). The data is generated from users' actions that are recorded automatically or manually logged through a computer or mobile phone. The emergence and rapid growth of social media platforms like Facebook (launched 2004), Twitter (launched 2006), and FourSquare (launched 2009) have contributed to the growth of applications "apps" that rely on user-generated location information. Combined with other variables, it is possible to quickly generate data-driven decisions. In other words, quantitative data is now used to support a variety of mundane and strategic decisions by everyday people.

Planners have traditionally gathered data about planning problems and issues and solicited feedback about planning proposals through formal and informal consultations. Some of this work is now conducted electronically by using social media platforms. A cottage industry of "app" developers now serve planning professionals[39]; they facilitate the collection and linking of user data with publicly available data to draw conclusions about people's behaviors and aspirational goals, thereby providing practical guidance for individualized decision-making. At the same time, planners should remember that urban management functions like crowd control are greatly facilitated by the range of new technologies and data streams that are available to planners and to law enforcement.[40]

2.7 Concluding Comments

American society is deeply polarized politically and socially. In part, this polarization is an outgrowth of great income and social inequality.[41] Poverty can no longer be an inner-city phenomenon because more than half of the country's poor people live in suburban environments (Kneebone and Berube 2013). Addressing suburban poverty will create new planning and design challenges for planning practitioners across the country. Planners must address the need for safe and affordable housing for home buyers and renters, the provision of adequate public transportation options for low-income people as well as social services like child care and professional

[39] Evans-Cowley, J. 2017. The Best Planning Apps for 2017, January 4, 2017. Available at: https://www.planetizen.com/node/90507/best-planning-apps-2017. Retrieved on January 16, 2017.

[40] Police to use LSE crowd control app, London School of Economics and Political Science, Available at: http://www.lse.ac.uk/website-archive/newsAndMedia/newsArchives/2012/07/crowd-control-app.aspx. Retrieved on March 7, 2017.

[41] Income Inequality. Available at: https://www.census.gov/topics/income-poverty/income-inequality.html. Retrieved on April 25, 2017

development services for those individuals who are reentering the workforce. These challenges may be further complicated because of suburban dwellers' real and perceived anxieties about crime. In many suburbs, the discussion about these anxieties is likely to evoke racial and ethnic tensions that must be carefully defused. At the same time, natural calamities, human-made disasters, as well as foreign and domestic terror attacks have become a part of contemporary life, and this trend is likely to continue for the foreseeable future. Threats can emerge from almost anywhere and manifest in many ways. Planners and practitioners must develop new skills to support civic education, emergency preparedness, response training, and planning in/for post-disaster situations.

References

Bacon E (1976) Design of cities. Penguin Books, Middlesex
Barnes T (2003) What's wrong with American regional science? A view from science studies. Can J Reg Sci 26(1):3–26
Bicknell J, Dodman D, Satterthwaite D (eds) (2009) Adapting cities to climate change: understanding and addressing the development challenges. Earthscan, London
Blakeley E, Snyder M (1997) Fortress America: gated communities in the United States. Brookings Institution Press, Washington, DC
Black A (1990) The Chicago Area Transportation Study: a case study of rational planning. J Plan Educ Res 10(1):27–37
Brown P (1942) Indian architecture: Islamic period. D. B. Taraporevala Sons & Co., Bombay
Bruegmann R (2005) Sprawl: a compact history. University of Chicago Press, Chicago
Campbell H, Masser I (1995) GIS and organizations. Taylor & Francis, London
Cisneros H, Engdahl L (eds) (2009) From despair to hope: hope VI and the new promise of public housing in America's cities. Brookings Institution Press, Washington, DC
Cole E (ed) (2002) The grammar of architecture. The Ivy Press, East Sussex
Davis K (1965) The urbanization of the human population. In: Scientific American (ed) cities: a scientific American book. Alfred A. Knopf, New York, pp 3–24
Downs A (2005) Smart growth: why we discuss it more than we do it. J Am Plan Assoc 71(5):367–378
Duany A, Plater-Zyberk E, Speck J (2000) Suburban nation: the rise of sprawl and the decline of the American dream. North Point Press, New York
Duany A, Plater-Zyberk E (2003) The lexicon of new urbanism. DPZ & Co., Miami. (Available from http://www.dpz.com/Books/All)
Federal Geographic Data Committee (1994) The 1994 plan for the national spatial data infrastructure: building the foundation or an information-based society. Federal Geographic Data Committee, Washington, DC
Friedmann J (1987) Planning in the public domain: from knowledge to action. Princeton University Press, Princeton
Gans H (1969) The levittowners: ways of life and politics in a new suburban community. Columbia University Press, New York
Garreau J (1991) Edge cities: life on the new frontier. Anchor Books, Random House Inc., New York
Gillham O (2002) The limitless city: a primer on the urban sprawl debate. Island Press, Washington, DC
Planning Commission (Government of India) (2013) The twelfth five year plan (2012–2017): faster, more inclusive and sustainable growth. Sage Publications, New Delhi

Gurda J (1999) The making of Milwaukee. The Milwaukee County Historical Society, Milwaukee
Hall P (1988) Cities of tomorrow: an intellectual history of urban planning and design in the twentieth century. Blackwell, Oxford
Hayden D (2003) Building suburbia: green fields and urban growth 1820–2000. Pantheon Books, Random House Inc., New York
Huxhold W (1991) An introduction to urban geographic information systems. Oxford University Press, New York
Ingram G, Hong Y (2009) Evaluating smart growth: state and local policy outcomes. Lincoln Institute of Land Policy, Cambridge, MA
Jackson K (1985) Crabgrass frontier: the suburbanization of the United States. Oxford University Press, New York
King M (1981) Chain of change: struggles for black community development. South End Press, Boston
Kitchen R, McArdle G (2016) What makes big data, big data? Exploring the ontological characteristics of 26 datasets. Big Data Soc 3(1):1–10
Knapp G, Talen E (2005) New urbanism and smart growth: a few words from the academy. Int Reg Sci Rev 28(2):107–118
Kneebone E, Berube A (2013) Confronting suburban poverty in America. Brookings Institution Press, Washington, DC
Knox P (ed) (2014) Atlas of cities. Princeton University Press, Princeton
Kostoff S (2010) A history of architecture: settings and rituals, 2nd edn. Oxford University Press, Oxford
Kostoff S (1991) The city shaped: urban patterns and meanings through history. Bullfinch Press/ Little, Brown and Company, Inc., Boston
Larson E (2003) The devil in the white city: murder, magic, and madness at the fair that changed America. Crown Publishers, Random House, New York
Lynch K (1981) Good city form. MIT Press, Boston
Low S (2003) Behind the gates: life, security, and the pursuit of happiness in fortress America. Routledge, New York
Miller D (1996) City of the century: the epic of Chicago and the making of America. Simon & Schuster, New York
Mollenkopf J (1983) The contested city. Princeton University Press, Princeton
Plummer A (undated, modified 2007) The Chicago Area Transportation Study: creating the first plan (1955–1962). Online resource, http://www.surveyarchive.org/Chicago/cats_1954-62.pdf. Last accessed January 2, 2017
Mumford L (1961) The city in history. Harcourt, Inc, New York
Ramasubramanian L (2009) Geographic information science and public participation. Springer, Heidelberg
Rapoport A (1969) House form and culture. Prentice Hall, Upper Saddle River
Rosenzweig C, Solecki W (eds) (2015) Building the knowledge base for climate resiliency: New York Panel on Climate Change 2015 report. Ann N Y Acad Sci 1336:1–150
Squires G (ed) (2002) Urban sprawl: causes, consequences, and policy responses. Urban Institute Press, Washington, DC
Stimpson C, Dixler E, Nelson M, Anthony K (1981) Women and the American city. University of Chicago Press, Chicago
Taylor N (1998) Urban planning theory since 1945. Sage, Thousand Oaks
Wachs M (1979) Transportation for the elderly: changing lifestyles, changing needs. University of California Press, Berkeley
Wu F (2015) Planning for growth: urban and regional planning in China. Routledge, New York

Chapter 3
Case Studies

3.1 Introduction

Our two case studies both represent "islands" with well-demarcated boundaries: in case of Roosevelt Island because it is indeed a true island in New York City's East River and in case of Hunts Point because it forms a peninsula that is separated from the rest of the Bronx by an imposing freeway (see Fig. 3.1). Located within the five boroughs of New York City, both case study areas have a long history of multiple phases of stable land use that are punctuated by dramatic changes (see Table 3.1). New York City has a long history of using islands for undesirable uses (Rikers, prison; Hart, cemetery; North Brother, quarantine hospital, then drug addiction treatment center; Randall, all previous uses and now a sewage treatment plant). Compared to that, our case study neighborhoods represent more common uses, especially during the last century.

In spite of the initial attraction of working with islands, data does often not conform to such natural boundaries. Census data is still quite often available at the census tract level, of which there are three in Hunts Point and two on Roosevelt Island. But virtually all other data holdings, from police precincts to neighborhood tabulation areas, community board, or school districts, capture a much larger area and do not allow for downscaling.

For both case studies, we have been relying on a multitude of local sources from blogs and Web pages to self-published monographs. We cannot stress enough how important it is to peruse such unofficial data. Facts need to be checked, of course, but it is relatively easy to learn what sources can be trusted. In particular, we would like to acknowledge the archival work of Paul DeRienzo, an independent investigative reporter who collated virtually all maps and photographs about Hunts Point that are accessible via the Internet.

This chapter is annotated with a good number of figures and photos but far fewer than fit into a printed volume. Our book website contains many more illustrations that describe the planning context for both case study areas.

L. Ramasubramanian, J. Albrecht, *Essential Methods for Planning Practitioners*,
The Urban Book Series, https://doi.org/10.1007/978-3-319-68041-5_3

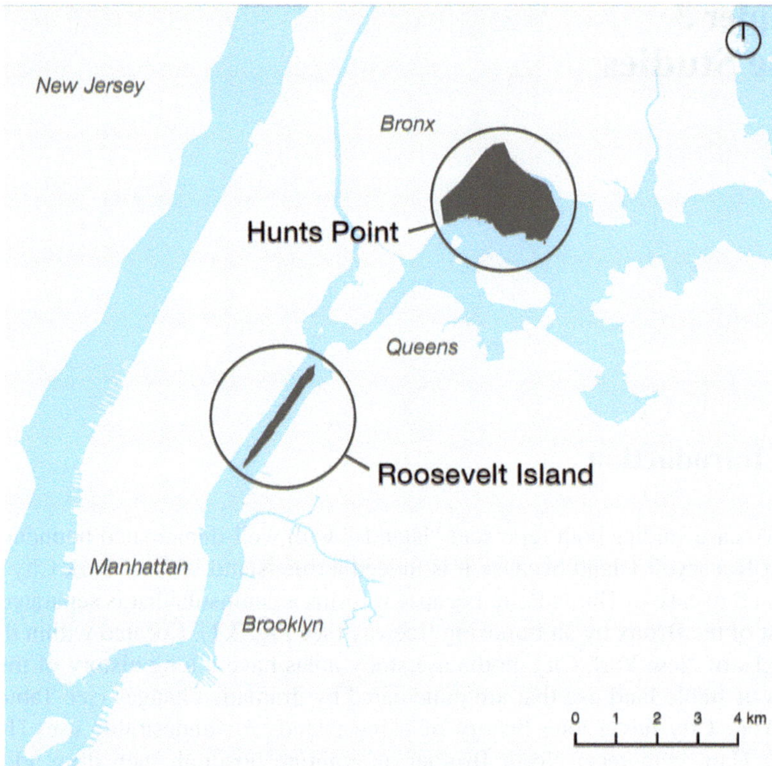

Fig. 3.1 Geographical setting of the two study areas

Table 3.1 Historical land use of the two case study areas

Phase	Hunts Point	Roosevelt Island
1650–1850	Farm land	Farm land
1850–1910	Mansions	Prison and mental asylum
1910–1970	Residential/industrial	Quarantine zone and hospital
1970–2010	Commercial/residential	Residential
2010–	Environmental resilience and regional service center	Residential/university/high tech industry

3.2 Hunts Point

The Hunts Point peninsula, our study area, is part of the South Bronx (Bronx Community District 2) and bounded by the East River and the Bronx River. The peninsula is isolated from the rest of the city by the Bruckner Expressway. Community District 2 extends further west for several blocks that includes a large swath of residential land use (see Fig. 3.2). The peninsula itself has a residential

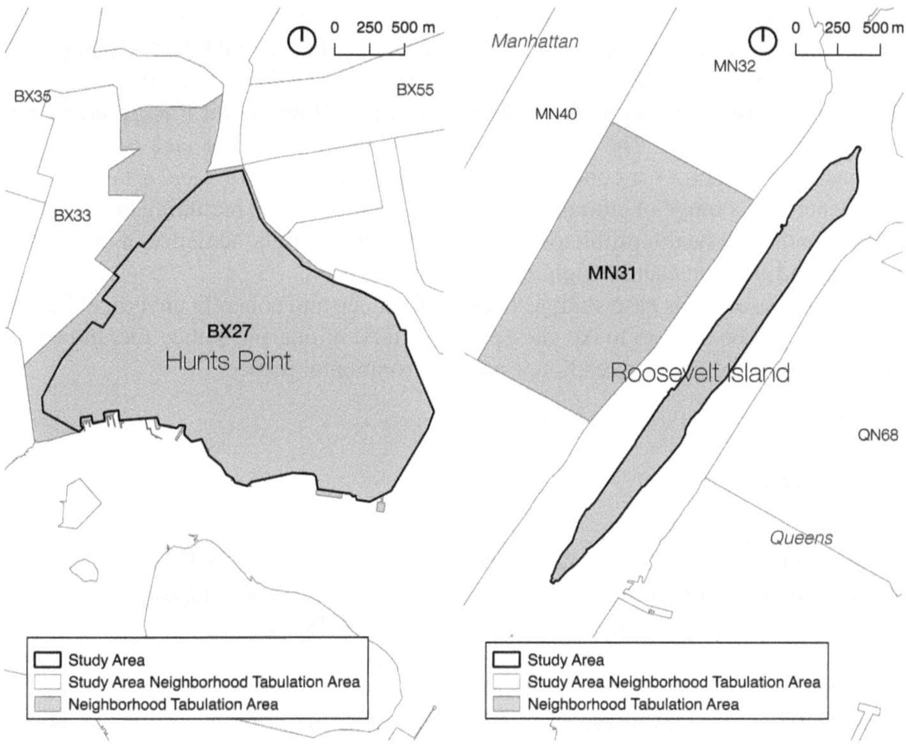

Fig. 3.2 The two study areas and their corresponding neighborhood boundaries

core, surrounded by a variety of industrial and manufacturing uses. Access to the waterfront is largely blocked by the Hunts Point Terminal Produce Market, the Hunts Point Meat Market, and the New Fulton Fish Market at Hunts Point. Collectively, these cooperative markets form the largest food distribution center in the world and occupy about half the land in the Hunts Point peninsula (about 320 acres). According to the NYC Economic Development Corporation (NYCEDC), the residential core of the Hunts Point peninsula is home to about 12,000 residents, although Community District 2 itself has a total population of 52,246 residents according to the 2010 census.[1]

Hunts Point, like many marginalized communities, has been overstudied. The 2005 Hunts Point Vision Plan identified (1) optimizing land use including creating a special district rezoning to expand industrial and large retail uses and creating/strengthening a buffer between residential and noxious industrial uses, (2) implementing workforce solutions, (3) creating connections including improving waterfront access and pedestrian safety improvements, and (4) improving traffic safety and efficiency for the approximately 15,000 trucks that access the food

[1] NYCDCP Bronx District 2 http://www.nyc.gov/html/dcp/html/neigh_info/bx02_info.shtml.

distribution center every day.[2] A portion of the peninsula was rezoned, and a special Hunts Point District was created in 2008. Recent studies include the DCP Sheridan Expressway Study, $20M in funding for demonstration projects related to resilience,[3] and some support for a new Metro North station at Hunts Point linking it to Penn Station in Manhattan.[4] The challenge in Hunts Point is not the lack of plans and visions but the lack of a coherent strategy *that puts neighborhood residents first.* Residents face a range of cumulative negative health impacts including high rates of asthma and other health problems, lack of green space, high incidence of pedestrian injuries and fatalities, and a high unemployment rate.

The purpose of this case study is to provide a deep and coherent analytical framework that can be used to make site-specific interventions and policy recommendations to foster community resilience and environmental justice.

3.2.1 History

Hunts Point is a peninsula located at the confluence of the Bronx River and the East River, which is actually a tidal strait connecting Upper New York Bay to the Long Island Sound. The total land area is approximately 690 acres (2.8 km^2). The study area was called Quinnahung (long high place) by the Lenape (American Indians) who settled here before the age of Henry Hudson's European exploration of the area. This name refers to the spine of an otherwise flood-prone peninsula, which now forms its residential core. The Indians sold the land in 1663, which was subdivided into a dozen farms known as the West Farms (west of the Bronx River). Original European settlements were right on the waterfront (The Grange) at the southernmost end of Hunts Point Avenue, Leggett's mansion at Oak Point. The first land holdings were rather large and transitioned between 1850 and 1900 into a home and vacation spot of New York City's elite with large country estates and meadow lands (see Fig. 3.3). The large houses were abandoned toward the end of the nineteenth century and were demolished one by one between 1900 and 1910, although Oak Point and Barretto Point remained meadows and a resort area with boat houses and an amusement park (see Fig. 3.5). Oak Point was bought in 1905 by a railway line for the establishment of a large railyard. Once a railway stop was built on Bruckner Boulevard, developers embarked on significant apartment building projects on both sides of the boulevard. A subway line followed suit in the 1920s, although most residential building activity remained to the northwest of our study area in what is known as Longwood – the other part of Bronx Community Board 2 (see Figs. 3.2 and 3.4).

[2] NYCEDC Hunts Point Peninsula http://www.nycedc.com/project/hunts-point-peninsula.

[3] Hunts Point Rebuild By Design Proposal http://www.rebuildbydesign.org/our-work/all-proposals/winning-projects/hunts-point-lifelines.

[4] Metro North Stations in Bronx http://www.nytimes.com/2014/01/09/nyregion/cuomo-supports-metro-northexpansion-in-the-bronx.html.

Fig. 3.3 Private land holders in 1868

Fig. 3.4 Sanborn map from 1921 depicting relationship among historic features

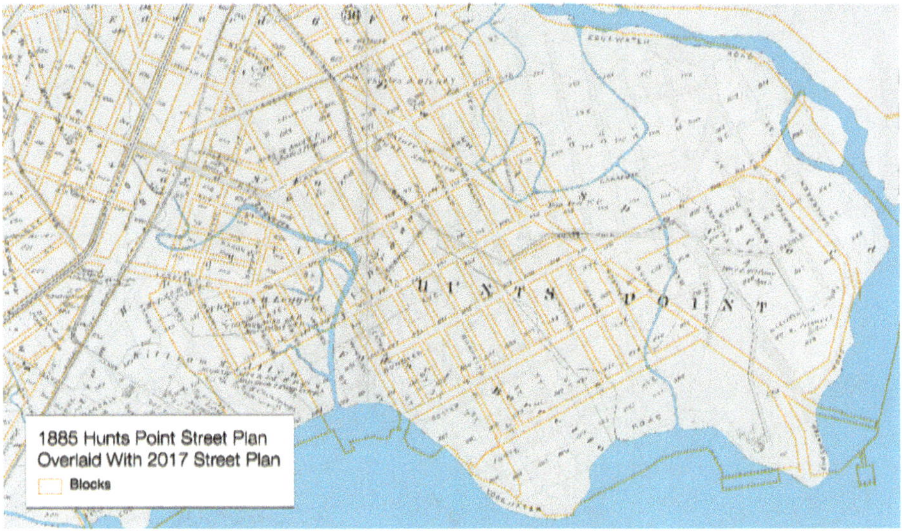

Fig. 3.5 Hunts Point plan from 1885 overlaid on current geography

The railway line also provided ready access for burgeoning industrial and commercial activity. The American Bank Note Company opened in 1911 and employed for many decades some 2000 people (Fig. 3.5). ConEd, a gas and electricity company, bought practically the eastern half of the peninsula and built a gas holder that remained in place until the late 1960s, when New York City's produce market moved from southern Manhattan to make use of the vicinity's access to railways, the freeway, and potentially sea-based transport (see Fig. 3.6). The location turned out to be perfect to combine several wholesale markets, prompting the regional meat market to move here in 1974 and the New Fulton Fish Market in 2005. Figure 3.7 shows the location and impact of the 670 businesses with some 13,785 employees in the commercial zone of Hunts Point that serve the metro New York region with an annual payroll of three fourths of a billion dollars and a total revenue of over $3 billion (US Census 2016). Approximately, 3800 trucks travel to the market each day, with many additional trucks also serving other enterprises in the vicinity, an important employment cluster (PlaNYC 2011).

While most of the land area in Hunts Point is dominated by industry (see Fig. 3.9), there is a small but dense residential pocket that occupies the high ground in the northern half of the peninsula bounded by Garrison Avenue and Randall Avenue on the north and south and Longfellow and Tiffany streets on the east and west (see Fig. 3.8). It consists primarily of pre-World War I apartment buildings with a smaller number of semidetached multiunit row houses. There is one public housing project on Hunts Point Avenue built in 1965 that contains 13 apartment buildings that were 4- and 5-stories high (Fig. 3.9).

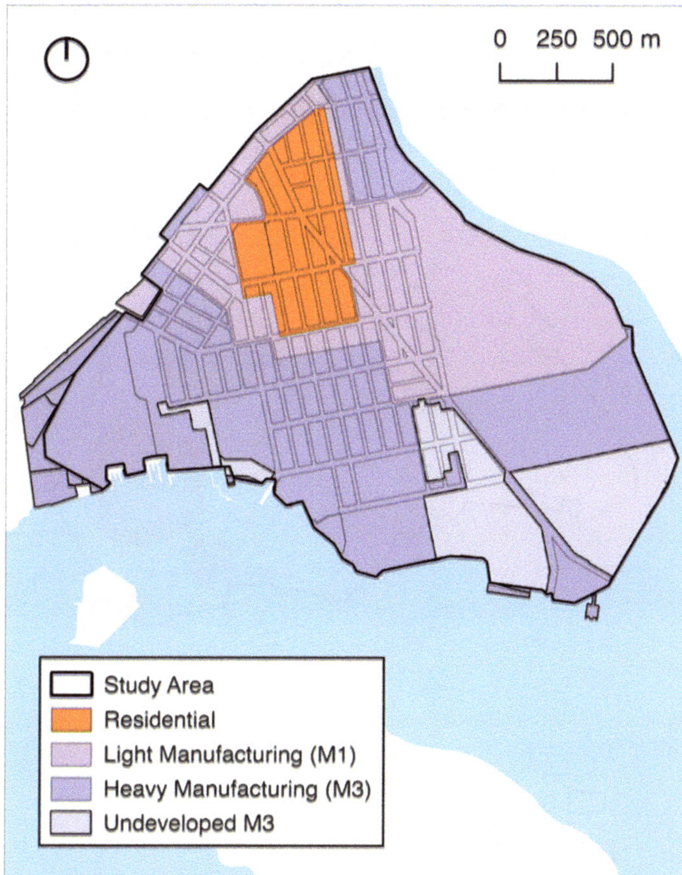

Fig. 3.6 Land use in Hunts Point in 1957

On the south shore of the study area lies an 800-bed jail barge to serve as overflow facility for medium to high security prisoners from Rikers Island just south of Hunts Point.

3.2.2 Demographics

Parallel to the historic phases described in the previous section, the demographics of Hunts Point changed in ethnicity and wealth. After the wealthy farmers and estate owners of British origin, the area saw an influx of middle-income Irish and Italian Catholics and central European Jews. Until the 1950s, the residential community on the peninsula was predominantly Jewish. The change in population began in the 1940s when Puerto Rican families began to leave East Harlem and by 1960 formed

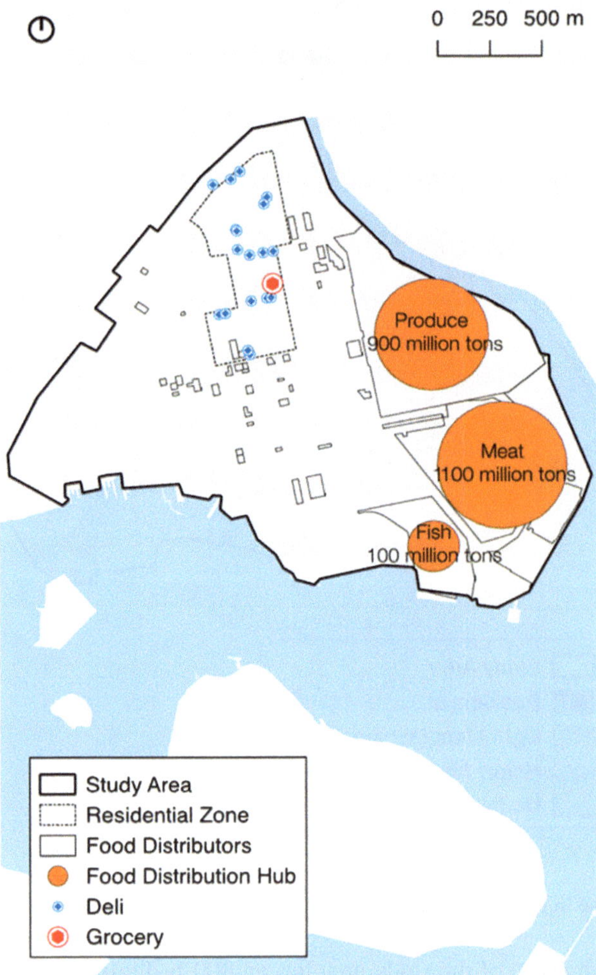

Fig. 3.7 The dominance of the regional food distribution centers in Hunts Point

the majority of the residents in Hunts Point. Related to that change, the relative wealth of the residents dropped from slightly above city-wide average to the lowest decile with an average adjusted gross income in 2014 of a mere $25,500. The number of residents, some 13,000 according to latest census figures, has not changed significantly over the past 100 years though. A third of the population now is foreign-born, with a significant number hailing from Dominica, Africa, and Central America. Altogether, the rate of neighborhood change has dropped significantly; some 67% of the residents lived at the same address 5 years ago.

Although there are approximately the same number of working-age people and jobs in Hunts Point, only about a quarter of the resident workforce has found

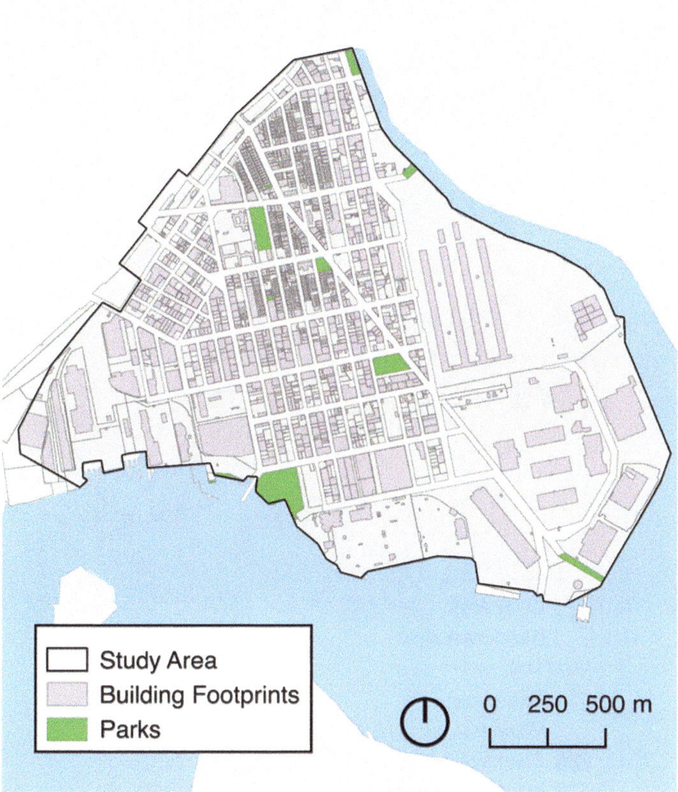

Fig. 3.8 Building foot prints at Hunts Point

employment within the case study area. Three quarters of the people working in Hunts Point hail from throughout the city and the neighboring Westchester County. On top of the significant burden of truck traffic to the wholesale markets (see Fig. 3.13), this compounds the traffic problem by adding another approximately 15,000 commuters, who are fairly equally divided into car and public transit passengers (see Fig. 3.10).

3.2.3 Community Burdens

In 2015 (the latest year for which figures were available at the time of writing), the unemployment rate in Hunts Point was almost 14%, compared to about 6% for New York City as a whole. Hunts Point is the worst-ranking NYC neighborhood with respect to opioid-involved hospitalizations (NY State 2015). It is also in the top

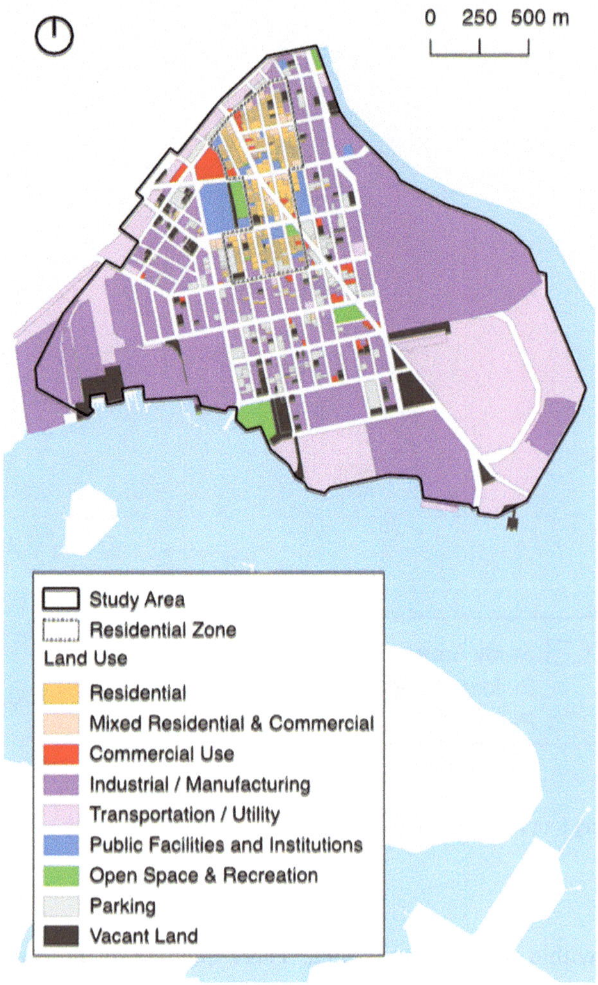

Fig. 3.9 Current land use in Hunts Points, NY

tier for STD infections, single mothers, and teen mothers (NY State 2015). The ratio of registered sex offender to residents is five times as high as in the city as a whole (NY State 2015). The overall crime rate, while below the national average is among the highest in New York City; although due to the relatively small residential community, these numbers are easily skewed. Unusual for the poorer neighborhoods in the Bronx, Hunts Point has very few community gardens; it does, however, have its own farmers market. Figure 3.11 depicts Hunts Point's poor ranking among New York City neighborhoods (calculations by the authors).

Fig. 3.10 Where Hunts Point residents commute to

3.2.4 Community Resources

It is important to note that every community, no matter how burdened it looks, also has indigenous resources. In case of Hunts Point, there are many community-based organizations (CBOs), each addressing a different type of community need. Back in the 1960s, New York City's Council Against Poverty (CAP) joined forces with federal programs to set up financially support community corporations. This purely economics-oriented organization was de facto replaced by an environmental justice-oriented organization called Sustainable South Bronx (SSBx). While SSBx is a

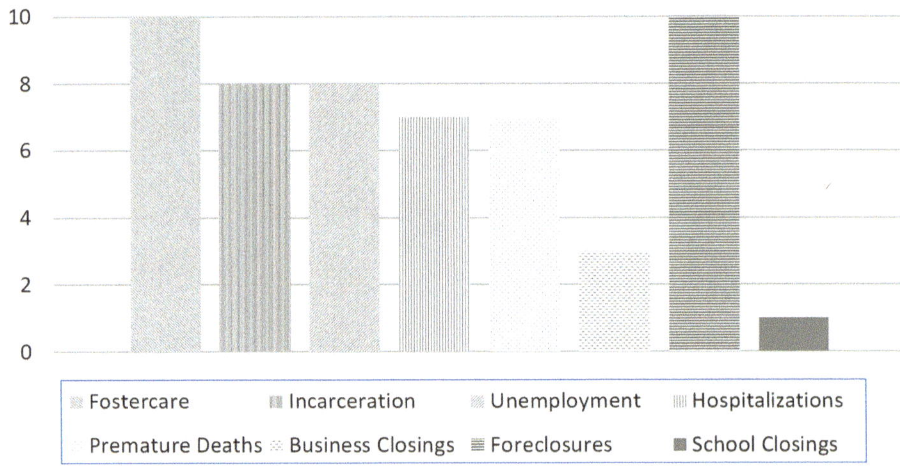

Fig. 3.11 Hunts Point's New York City-wide rank for selected community losses. The value 10 stands for the tenth or worst decile; so, for example, school closings are not an issue in Hunts Point but foster care and foreclosures are

Fig. 3.12 Community organizing activities (Image courtesy: Rebuild by Design)

conventional nonprofit organization, the City has been sponsoring considerable outreach efforts as part of its larger redevelopment initiatives surrounding the fish, meat, and produce distribution centers. The City's economic development corporation is financing a neighborhood outreach team,[5] which in turn runs a series of community workshops and community building activities like an annual summer festival (see

[5] Seehttps://medium.com/hunts-point-resiliency/were-hiring-join-our-neighborhood-outreach-team-b29fc72a8f76?mc_cid=075886a680&mc_eid=%5BUNIQID%5D.

Fig. 3.12) and a blog called Connection Stories.[6] Together with local schools, and their parent-teacher associations, every residential neighborhood will always have some resources that can be tapped in order to turn planning challenges into planning opportunities. The following are organizations identified as key stakeholders in this neighborhood of a mere 12,000 residents: the BLK Projek, GRID Alternatives, Hunts Point Economic Development Corporation, Mothers on the Move, The Point Community Development Corporation, Rocking the Boat, SoBro, Sustainable South Bronx, Urban Health Plan, and Youth Ministries for Peace and Justice.

3.2.5 Planning Challenges

3.2.5.1 Nuisance Land Uses

It is evident from the introduction that the Hunts Point area has been accumulating a number of NIMBY land uses, from smelly fish to the world's largest floating prison. For example, Fig. 3.14 shows the distribution of noisy and smelly waste management sites. While it could be a point of discussion whether these best out of one's sight facilities are indeed best to be concentrated in a few isolated places, the matter of fact is that Hunts Point is home to some 13,000 residents who are clearly suffering from a concentration of environmental justice issues.

3.2.5.2 Truck Traffic

The Hunts Point Food Distribution Center generates approximately 27,400 tons of waste per year, roughly 75% of which is organic and all of which is being hauled away in trucks for disposal (see Figs. 3.13, 3.14, and 3.15).

Another attractor for a large number of trucks is the cluster of waste transfer stations, prompting the New York Metropolitan Transportation Council to conduct a feasibility study for truck ferry access to the Hunts Point fish market.

3.2.5.3 Declining Residential Quality

While the residential cluster has been very stable over the past 100 years, the housing stock has started to deteriorate, and the residents willing to put up with poor housing and the environmental nuisances have become increasingly poorer. Arguably, there has been a long-term disinvestment in the housing stock as well as in amenities for residents. Hunts Point has a dearth of green space. Figure 3.16 shows the small percentage and isolation of green areas, some of which, like the

[6] See https://medium.com/hunts-point-resiliency/hunts-point-resiliency-connection-stories-85635940b662.

Fig. 3.13 Idling refuse trucks lined up in Hunts Point (Image courtesy: Hunts Point Studio at Hunter College)

Fig. 3.14 Hunts Point is a center for waste management

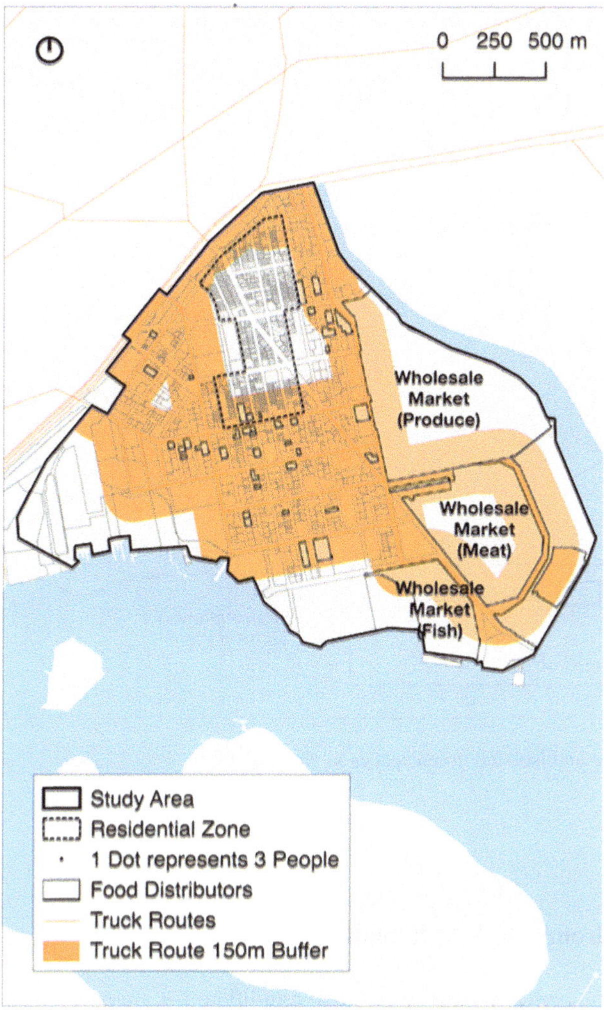

Fig. 3.15 Truck route buffers cover almost all of the study area

playground immediately west of the residential area, are not actually green but bare courts.

Due to the high level of poverty, many residents of Hunts Point are going hungry and malnutrition is a constant threat. There lies some considerable irony in the fact that right next to heaps of food for the region, the local population is served by only a single grocery store and a number of delis (the term delicatessen is a New York City euphemism for small neighborhood-based outlets that often serve limited healthy food options in resource-poor neighborhoods). Figures 3.18 and 3.19 juxtapose the spread of food distributors with the scarcity of local retail.

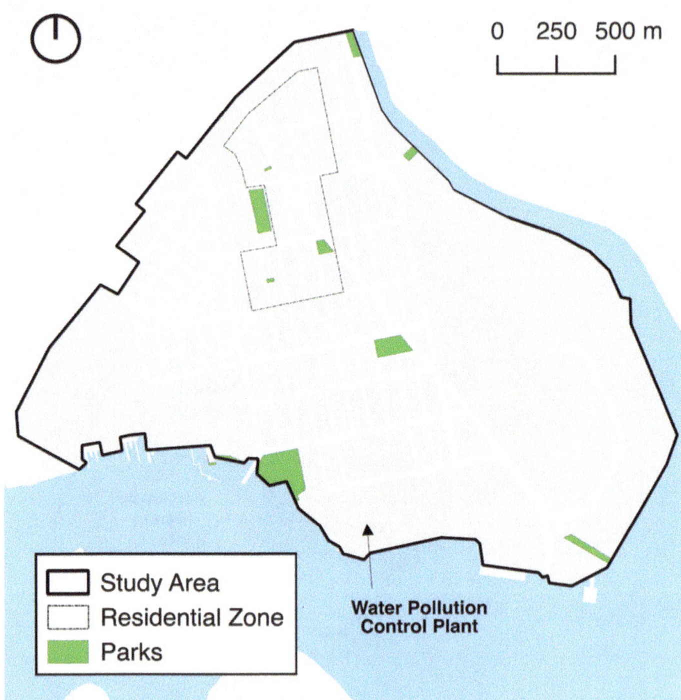

Fig. 3.16 Scarce and isolated green spaces in Hunts Point

3.2.5.4 Environmental Sustainability

The distribution center would be an ideal candidate for an *on-site* organics recovery operation. Such a facility could lower waste disposal costs, generate a clean source of energy, reduce truck traffic and related impacts both locally and regionally, decrease congestion, and reduce air pollution.

3.2.5.5 Design for Climate Resilience

As can be seen in Figs. 3.4 and 3.5, much of Hunts Point used to be a tidal swamp that is still prone to flooding, especially in light of predicted sea level rises due to global climate change. Figure 3.17 depicts the impact that various levels of inundation have on the area. While the residential core in the north will remain unaffected, virtually the whole economic base of the study area will face an increasing number of shutdowns due to flooding.

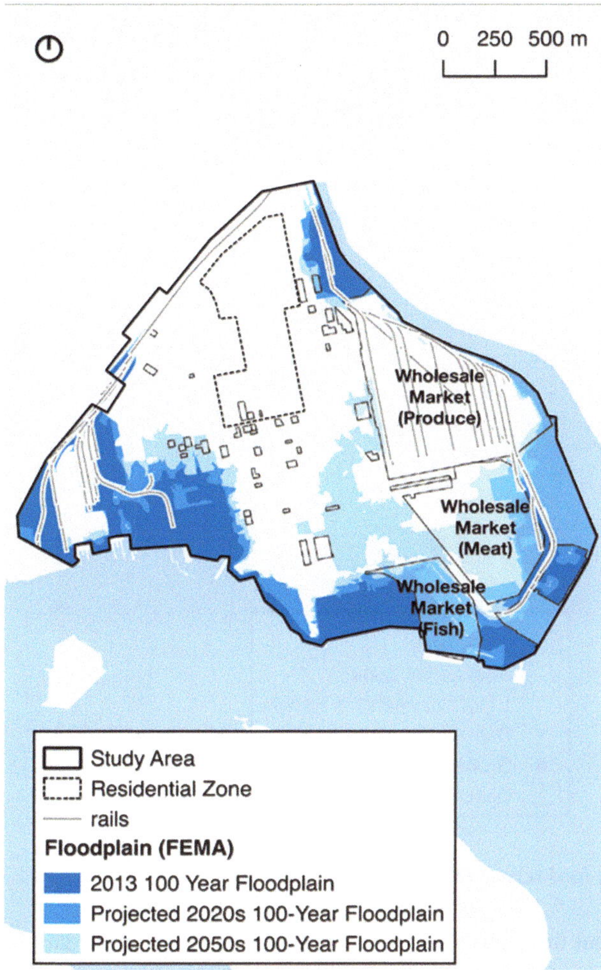

Fig. 3.17 Hunts Point is increasingly prone to be flooded

3.2.6 Planning Opportunities

3.2.6.1 Field Observations and Community Conversations

In Sect. 3.2.4, we pointed to the community resources that every planning professional should capitalize on. In Chapter 4, we are going to discuss methods that assist in envisioning the future. While we encourage every planner to explore her study area personally and on foot, it is important to connect with and engage local residents. The first step is a series of field observations, both by the planning

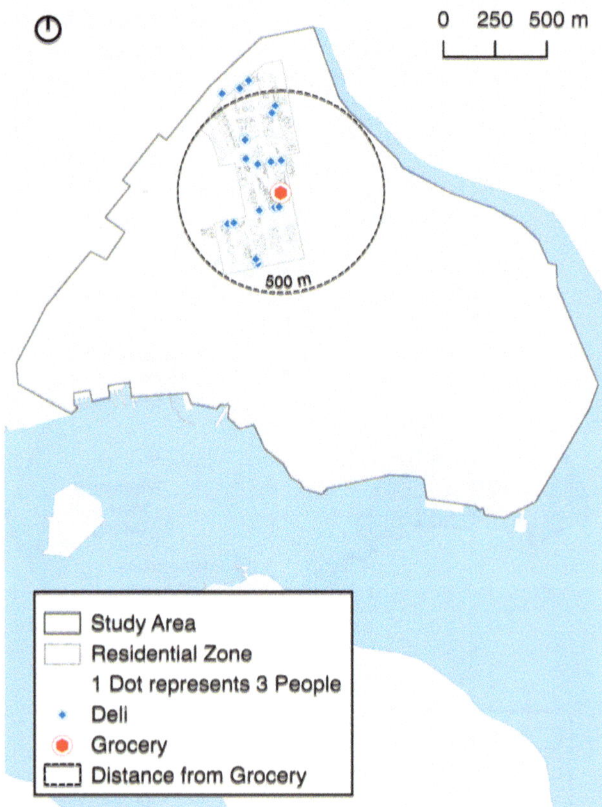

Fig. 3.18 Local food-related retail

professional but then enriched by local residents. An example can be found on this WikiMaps page,[7] where residents map the places they like or dislike (see Fig. 3.20).

This form of primary data collection forms the basis for community conversations (see Fig. 3.21), which in turn help to build trust. In the course of such reconnaissance, the planning professional will learn about all stakeholders and would be well advised to engage them all. This process has been mirrored by the authors, one of whom directed a planning studio for graduate students at Hunter College. The introduction to this case study described the goal of the studio to provide both policy recommendations and specific urban design interventions to help foster community resilience and address environmental and economic injustice in this community. As is common in real-world planning projects, such goals morph as the project progresses and new insights are gained by all participants. In case of the studio project, one tangible outcome was a set of recommendations to improve accessibility to green spaces and thereby improve the living conditions of the residents.

[7] See http://wikimapping.com/wikimap/HuntsPointCommunityMap.html.

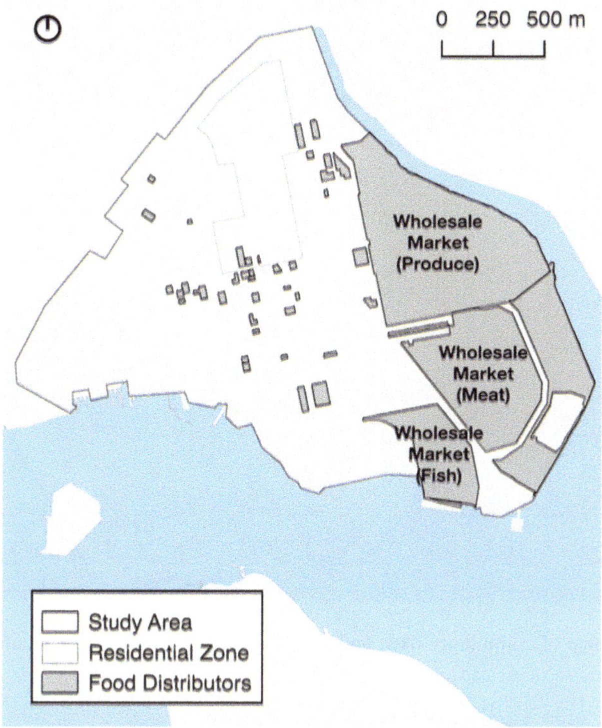

Fig. 3.19 Food distributors in Hunts Point

3.2.6.2 Improving Accessibility

If Hunts Point were in Manhattan or Brooklyn, then the large amount of waterfront would suggest a real estate boom. Similarly, most of the shoreline of neighboring Manhattan either has greenways or is in the process of building them, even in light of prior highway construction along what was perceived to be peripheral areas in the 1960s. One major initiative that is supported by many of the CBOs listed in Sect. 3.2.4 is the greening of the waterfront area. This still leaves the problem that these new potential recreational areas are if not far away then difficult to access by the residents in the northern part of Hunts Point (see Fig. 3.22). The closest access to small slivers of green space along the waterfront is more than half a mile away and requires facing relentless truck traffic.

The conflict between demands of residential versus commercial land use is not going to be decided in favor of one or the other. The solution space will therefore have to improve the living conditions of existing and future residents (the City just announced project called The Peninsula, a mixed use "campus" with some 740 units of affordable housing replacing the juvenile detention center on Spofford Ave), while separating and minimizing the effects of the distribution centers. One way of

Fig. 3.20 Wikimap of community likes and dislikes (Image courtesy: The Hunts Point Studio at Hunter College)

Fig. 3.21 A typical community consultation (Image courtesy: Hunts Point Studio at Hunter College)

doing that is to provide a free shuttle circulating from the residential area to Barretto Point Park. A temporary trolley that is currently in use during the annual Fish Parade serves as an example.

As we will see in the following case study on Roosevelt Island, the term "accessibility" has a wide range of connotations. Hunts Point residents, in their asset mapping exercise, criticized the lack of fresh food in the neighborhood. The farmers market, financially supported by the City, is supposed to ameliorate this but an obvious further reaching option would be the development of community gardens, similar to the developments in many other parts of the Bronx. One of the reasons this has not occurred yet is the unknown quality of the ground, given that much of Hunts Point is effectively a brownfield. Raised beds and the application of some urban farming techniques would go a long way to overcome the problems of disconnected youth in Hunts Point (CSS 2008).

3.3 Roosevelt Island

Roosevelt Island, located on the East River (the branch of the Hudson River that separates Manhattan from Queens on Long Island), is approximately 2 miles long and 800 feet wide (see Figs. 3.2 and 3.23). Commercial and residential buildings are concentrated in the center of the island. For many decades, the Coler Hospital campus occupied the northern and the Goldwater hospital the southern tip of the island. Although the Queensboro Bridge (connecting Manhattan with Queens) crosses the island, there is no access to the island from that bridge. The only other bridge, the Roosevelt Island Bridge, provides access to Queens. The island is currently served by a subway stop and a tram. Roosevelt Island is owned by New York City, but the Urban Development Corporation (UDC) took a 99-year lease on the island in 1969. The island has been a residential area only since the early 1970s, when the UDC created a planned community here. The island is governed by the Roosevelt Island Operating Corporation (RIOC).

The area of the former Goldwater hospital is being redeveloped to become an applied science and engineering campus (see Sect. 3.3.6). Further north, there are seven residential building complexes, a public library, and two schools, a public elementary school, and a private school for children with developmental disabilities. There are six historical landmarks on the island: the Smallpox Hospital (Renwick Ruins), the Strecker Memorial Laboratory, Blackwell House, the Chapel of the Good Shepherd, the Octagon Tower, and the Lighthouse. The local community is served by one Catholic church, several protestant churches, and one Jewish synagogue. The residential core is complemented by a large parking garage, Motorgate, and a small group of retail stores, which enjoy a cornered market on the island. There is one supermarket, which has a restrictive lease which stipulates that no other supermarkets be permitted on the island, a fact that many residents are dissatisfied with. There is, however, a popular farmer's market every Saturday, located beneath the Roosevelt Island Bridge access.

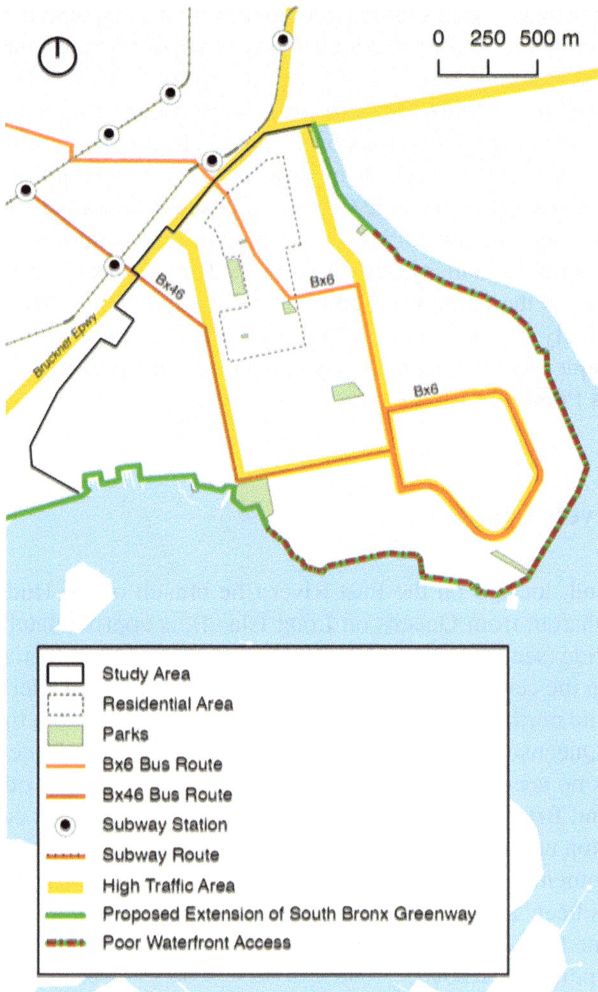

Fig. 3.22 Improving accessibility within Hunts Point

3.3.1 History

Formerly known as Blackwell's Island, it was a Blackwell family property until 1828 when it was sold to the City. Until the late 1960s, 90 of the total land area of 147 acres were unused and with the exception of a few caretakers, the majority of the then 3400 inhabitants were chronic care patients. In 1968, Mayor John Lindsay put together a committee to develop Welfare Island. Ed Logue, who was president of the UDC, chose architects Philip Johnson and John Burgee to produce a General Development Plan, completed in 1969, a modernist design calling for some 5000 units. Ed Logue's vision was to create a mixed-income community (30% low

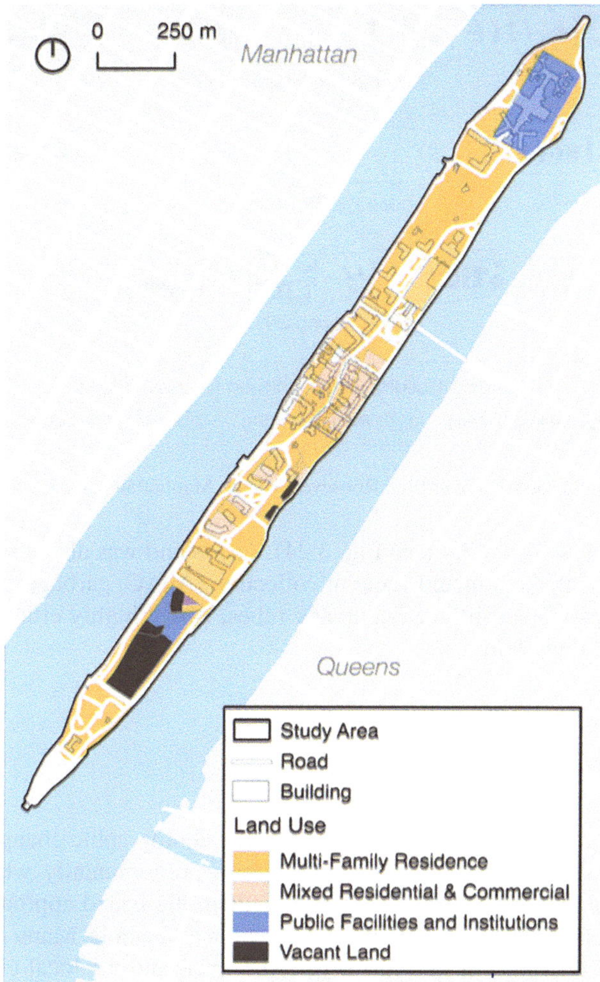

Fig. 3.23 Land use on Roosevelt Island Roosevelt Island, Manhattan:history

income, 45% middle income, and 25% luxury apartments), with special attention on providing housing for the elderly, disabled, and for employees of the hospital campuses on the island. Subsidized housing faced Queens, while market rate housing faced Manhattan. Roosevelt Island was originally designed to be pedestrian, served by an electric railway. Vehicle access was to be strictly limited to facilitate safe bicycle and pedestrian access for families. Private cars would be left at a 2000-car garage near the bridge to Queens. The community was planned to be self-contained, providing an elementary school, grocery store, cafes, post office, and other standard municipal services, as well as two pools, and a 300-room hotel. Logue also wanted his community to include plenty of open spaces for family sports and recreation, and community vegetable gardens, which were an unusual feature at the time in

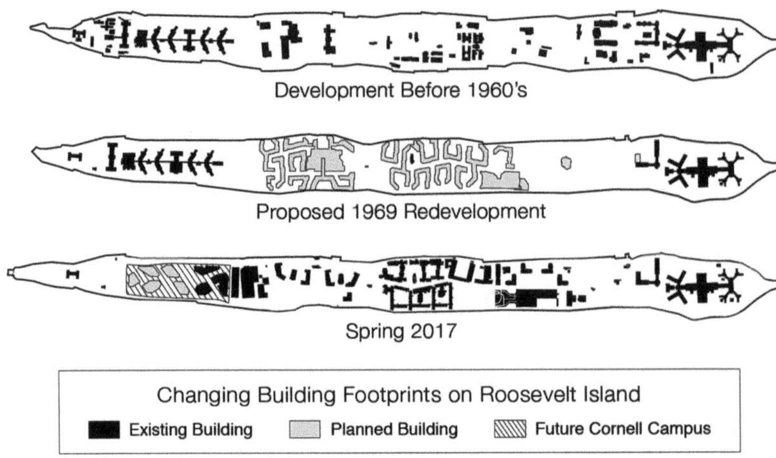

Fig. 3.24 Building footprints over time Roosevelt Island, Manhattan:history

overcrowded New York City (see Fig. 3.24). The island was also a site for innovative technology: the automated vacuum collection (AVAC) garbage control system, which functions via subsurface pneumatic tubes, is still highly efficient and is still studied around the world.

3.3.2 Demographics

Roosevelt Island has undergone some significant demographic changes. During its days as Blackwell's and Welfare Island, it was predominantly white. After the implementation of the General Development Plan, the Island approached a demographic mix comparable to the metropolitan region as a whole. Many residents work for the United Nations, located just across the river, and the local elementary and middle school contains a high proportion of international students. With approximately $68,000 in 2016, the median income is about 30% higher than that of New York City. The unemployment rate of 3.5% is considerably below New York City average. The United Nations Development Corporation issued a report in 1989 that led to an amendment in the Roosevelt Island General Development Plan, opening the door for higher income developments like The Octagon and Riverwalk. This second building boom between 2000 and 2009 (with some 1887 new units constructed during this period) resulted in a significant shift toward higher income occupants. Eastwood, designed specifically to house disabled, elderly, and hospital employees, has left the Mitchell-Lama Housing Program (a New York affordable housing program that funded over 100,000 units between the mid-1950 and 1970s) and is privatizing its units as current residents vacate. Three other subsidized buildings are moving toward privatization as well, although current residents are still protected (Table 3.2).

Table 3.2 Roosevelt Island in US Census Figures

Year	1910	1920	1930	1950	1960	1980	1990	2000	2010	2013
Total population	6990	5378	7591	5424	3626	6960	8190	9520	11,661	11,783
Housing units							2894	3252	4328	4056
Elderly					54.1%	19.6%	20%	17.7%	15.0%	18.7%
In poverty							20%	13.2%	12.2%	9.4%
Poor elderly							3%	1.7%	3.8%	1.9%
Rent burdened							18%	42.2%	34.2%	32.7%
Disabled								26.8%	N/A	14.3%
White	96.2%	95.3%	87.9%	86.4%	77.4%	71.6%	62.8%	49.2%	51%	53.1%
African-American	3.7%	4.7%	11.6%	12.8%	22.6%	20.5%	26.5%	27.2%	21.3%	14.4%
Asian	0.1%		0.5%	0.9%		3.8%	6.4%	10.7%	18.2%	24.6%
Other						3.1%	4.3%	12.9%	9.5%	7.9%
Hispanic								14.4%	14.9%	14.6%

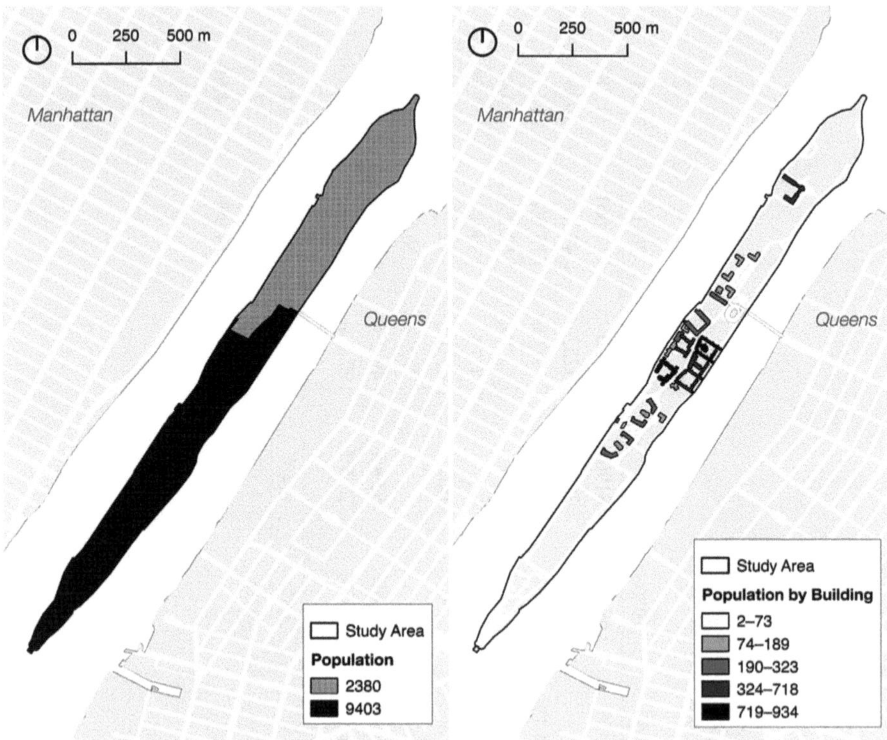

Fig. 3.25 Population map of Roosevelt Island

Together with the new Cornell University campus and its associated faculty housing, this will undoubtedly lead to a gentrification of the island (see Figs. 3.24 and 3.25).

3.3.3 Community Burdens

Roosevelt Island is served by several options for public transit. The Red Bus service consists of hybrid-electric buses that shuttle residents to on-island locations. They are ADA accessible and free to use. The MTA Q102 bus line connects Roosevelt Island with Astoria in Queens. The Roosevelt Island Tramway was constructed in 1976 along the Queensboro Bridge and provides access to Manhattan. In 1989, the New York City Subway created a station on Roosevelt along the F line. The Roosevelt Island subway station is one of the deepest in the metropolitan area, lying 100 feet underground (see Figs. 3.26 and 3.27).

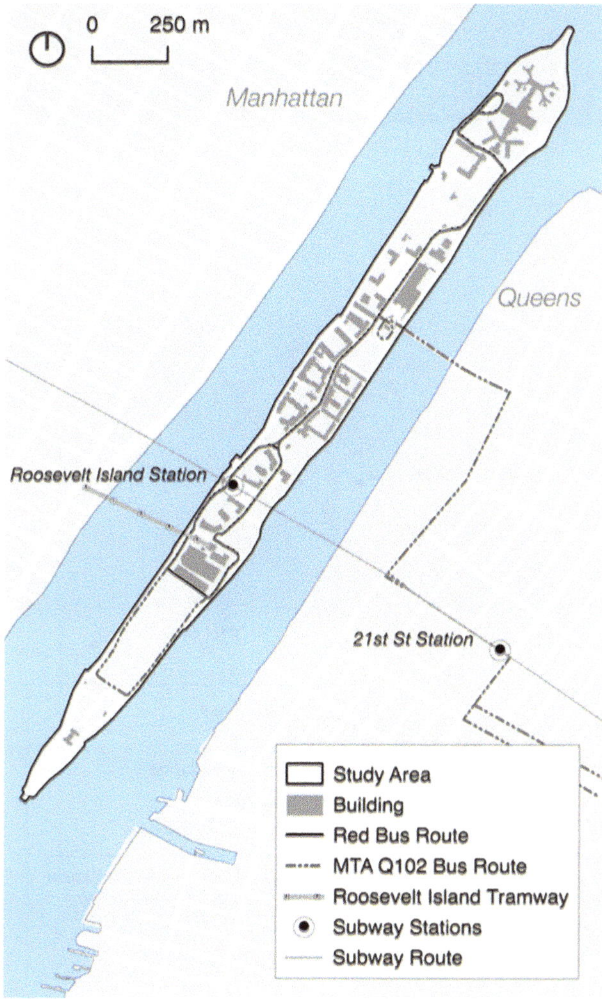

Fig. 3.26 Transit options

Pedestrian access is in high demand and has been increasingly marginalized by vehicular circulation. Additionally, many sidewalks and portions of the promenade do not meet ADA requirements, a fact that merits special attention in the Roosevelt Island community. Residents do not find Main Street as pleasant to walk due to lack of greenery and to the dark, canyon-like quality created by the high street side buildings. Bicycle circulation is impeded along the northern Queens-side promenade.

Some 75% of residents commute by subway. Figure 3.28 shows that most residents have a relative short commute to work, mostly in midtown Manhattan.

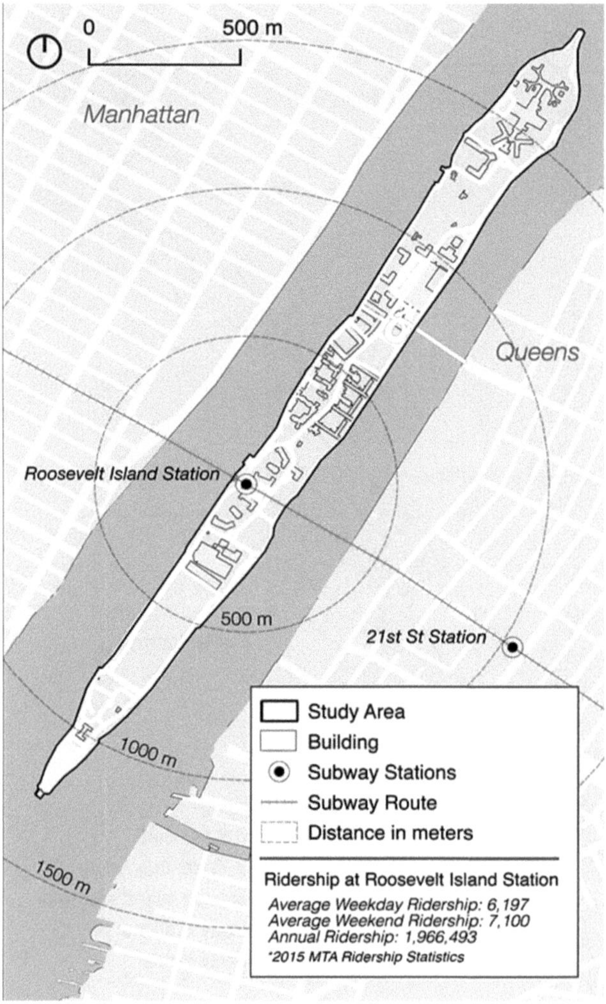

Fig. 3.27 Distances from Roosevelt Island subway station

3.3.4 Community Resources

Roosevelt Island is a very safe space for pedestrians and bicyclists. During a 4-year study period by the NYC Department of Transportation, no accidents for those two modes of transportation were recorded. A total of 5 intersections (out of a total of 24 on the Island) experienced vehicular accidents resulting in 17 injuries but no fatality.

Roosevelt Island is a comparatively healthy neighborhood by New York City standards. Figure 3.29 depicts the island community as being at the lowest needs rank, measured in terms of (i) poverty, (ii) unemployment, (iii) median household

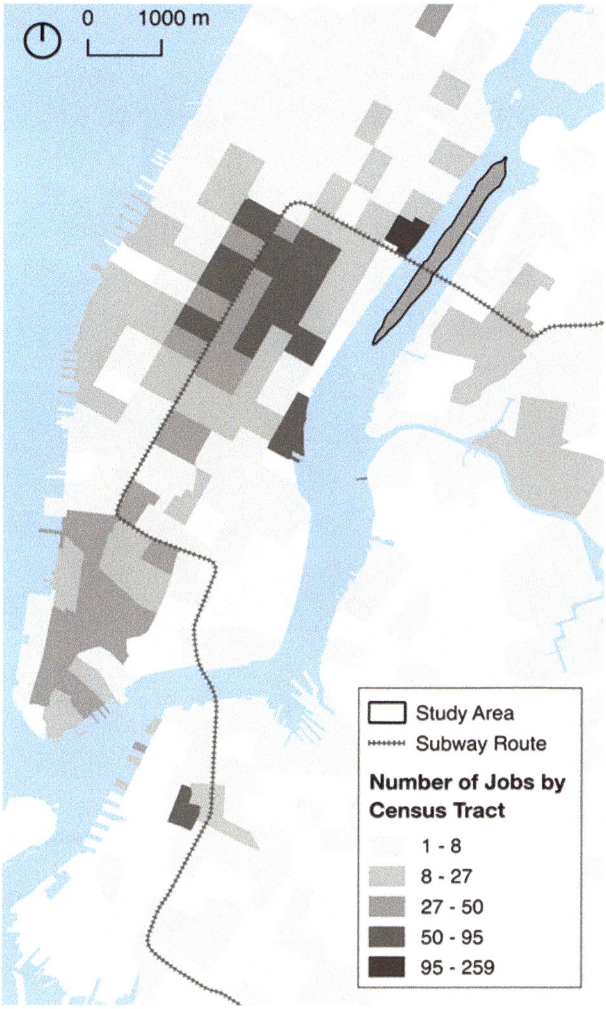

Fig. 3.28 Places of work for Roosevelt Island residents

income, (iv) educational attainment, and (v) infant mortality rate. Sheltered in the middle of a river and with relatively small amounts of vehicular traffic, Roosevelt Island seems like an oasis in the noise of New York City (See Fig. 3.30).

The average monthly rent on the Island is approximately $2200 – a bargain when compared to the equally small apartments on Manhattan's Upper East Side just a stone throw away. This level of affordability is due to the large percentage of rent-stabilized housing (see Demographics section above), resulting in the lowest bracket of rent as percentage of income in the whole city. RIOC also is a good landlord: no other NYC neighborhood has a lower 311 building maintenance

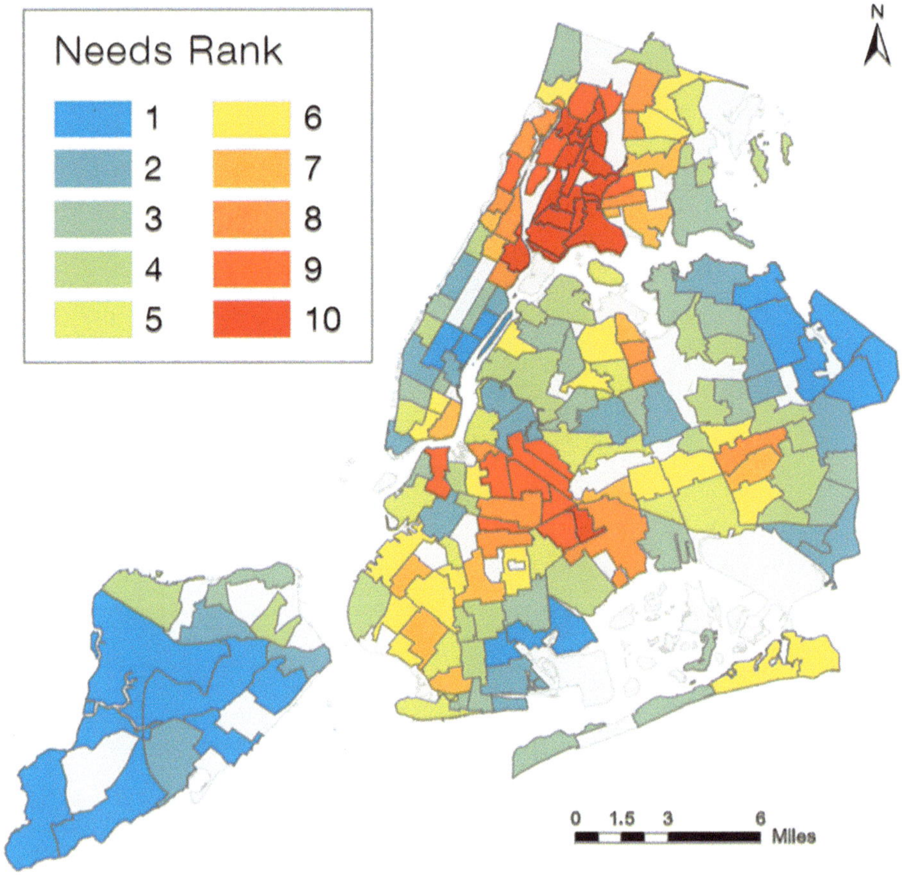

This map is based on the census tract-based aggregation of ranks across six variables:

I: % below poverty level, % unemployment, and median HH income
II: % with less than 9th grade eduaction and % with college education
III: infant mortality rate

For each variable, we created a scale of ten classes and then averaged to create a rank level for each of the above three categories. Each of the above three categories were then summed up and rank-ordered. The needs rank is the summary measure for the above procedure.

Fig. 3.29 Economic, education, and health needs in New York City neighborhoods

complaint volume. As a matter of fact, during the 7 years of recorded data, per capita call volume to 311 has been only 1/10 that of New York City overall, whereas back in the Hunts Point case study, the per capita call volume was 60% higher than in New York City.

In spite of resident's complaints about the scarcity of shopping options, Roosevelt Island ranks high in regard to per capita business openings (see Fig. 3.31).

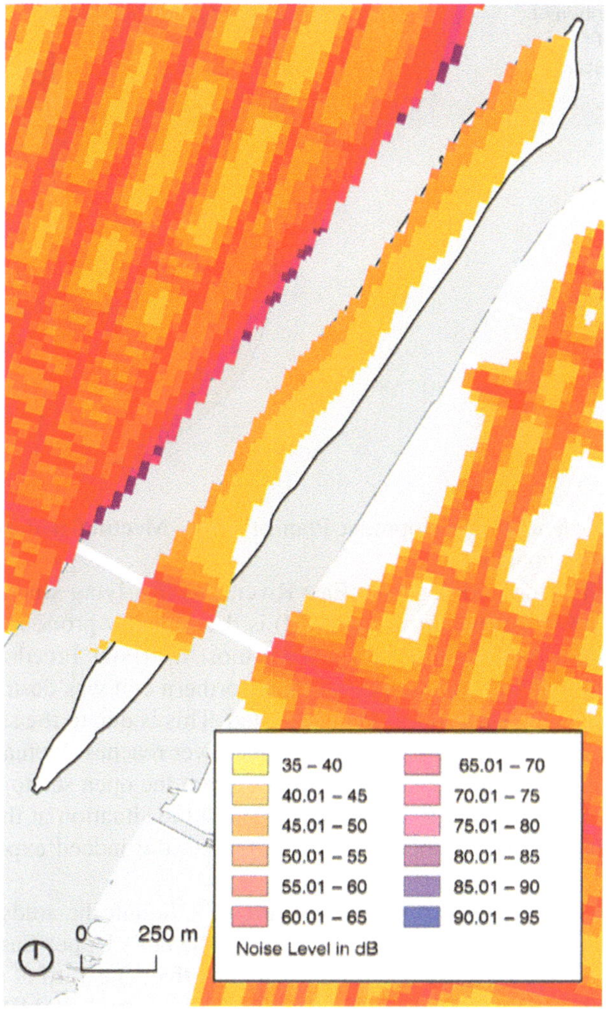

Fig. 3.30 Noise levels on Roosevelt Island (the Queensboro Bridge is clearly visible)

3.3.5 Planning Challenges

A unique planning challenge of Roosevelt Island is its political structure. Being on lease to RIOC is somewhat similar, though not the same as what planners encounter with gated communities elsewhere in the United States. Political representation is limited to the State capital in Albany – by New York City standards almost as far away as Canada. The City is expected to provide services but the planning rules of the City do not apply. Residents feel left out of many planning decisions, which is exacerbated by the fact that they often moved here because of the progressive

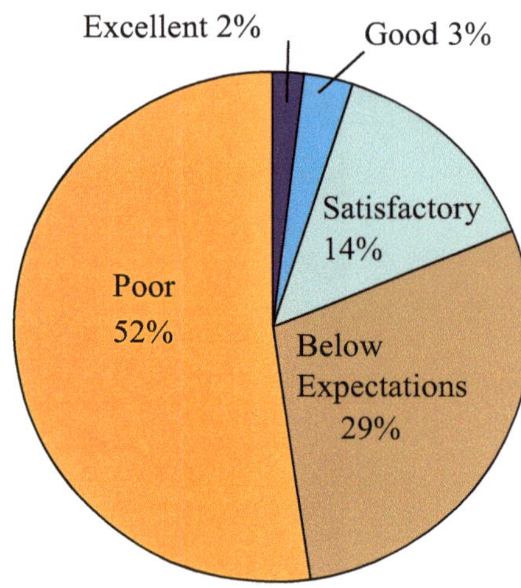

Fig. 3.31 Perception of overall quality of commercial services

underlying the General Development Plan of 1969. Meeting with the public can therefore be quite contentious.

As an island in the middle of the East River, this low-lying study area (90% of the island is 10 feet or less above sea level) is theoretically prone to flooding. Yet, during superstorm Sandy, only the southern-most tip (Four Freedoms Park) was flooded. In addition, one ConEd pump at the northern end was destroyed but altogether, the damage was much less than expected. This is due to the sheer discharge volume of the Hudson River, whose level in its lower reaches fluctuates only minimally; yet Roosevelt Island is far enough away from the open sea to be beyond the reach of the storm swell there. This contrasts with the situation at the Hunts Point study area, which, lying on the Long Island Sound, did indeed experience higher flood levels.

Locally, there are no sources of air pollution, although the study area is close enough to old and polluting heating units on Manhattan's Upper East Side as well as to the stacks of the Ravenswood power plant less than one third of a mile away in Queens to be affected by either easterly or westerly wind. The 2480 MW Ravenswood station is the state's most excessive carbon polluter. Independent of this regional supplier, an innovative project to supply emission-free electricity to the Island using tidal energy is expected to be expanded further.

On the demographic side, Roosevelt Island is rejuvenating. Being as close to densely populated Manhattan as one could possibly get, residents are used to and even welcome high-density, urban environments, especially since the housing stock itself is relatively new and hence easy to maintain. At the same time, some residents are concerned about the loss of intimacy, diversity, and its unique social history that contributed to making Roosevelt Island an attractive place to live. There is a strong sense of pride in the green spaces that the island provides, and the availability of waterfront views, even from within a high-rise unit, only adds to the attraction of the study area. Given the relative youth of the facilities, the infrastructure is in good shape, as evidenced by the low call volume to 311 (NYC 311 2017). The main issue, similar to the Hunts Point study area, although for very different reasons, is the question of transportation access.

The authors conducted a planning studio with students from the Hunter College Department for Urban Affairs and Planning in 2009. In spite of the subway stop, the tram to Manhattan, and the bus line to Queens, residents felt (and feel) cut off by the lack of a fixed bridge to Manhattan. In addition, they have expressed frustration with Main Street and its rather small town feel. The studio identified four categories of access-related issues: *placemaking*, *revitalization*, *infrastructure*, and *governance*.

As mentioned above, the islanders have a strong sense of place but are dissatisfied with the visual character of their community. Landscape elements from improved seating and greenspaces to orientation aids in the labyrinth of the building footprints (see Fig. 3.24) are relatively easy ways to address this issue. Revitalization pertains mainly to the drab look of Main Street (see Fig. 3.32). While it was mentioned above that Roosevelt Island enjoys a surprisingly large number of business openings, this can also be interpreted as a large amount of turnover – few businesses survive. The island's landlord is too bureaucratic and aloof to facilitate a market-oriented exchange of supply and demand of services. The infrastructure of the island was built for no more than 10,000 residents and is operating at or beyond capacity. Traditional forms of expansion by spreading into the neighborhood are not possible on an island, so improvements have to occur in situ. The governance problems are

Fig. 3.32 Drab view of Main Street (Image courtesy: Roosevelt Island Studio at Hunter College)

the hardest to overcome, especially as traditional forms of improving transparency and community involvement are either not used (311 system) or difficult to formalize in the context of traditional local representation. A prime example for the latter is the lack of participation in the planning process of the Cornell campus discussed in the following section. It does not help that the territory of the state senator representing Roosevelt Island encompasses geographically, socially, and ethnically completely different parts of New York City, namely, Spanish Harlem and the South Bronx.

3.3.6 Planning Opportunities

3.3.6.1 Transportation Improvements

The extended shape of the island causes some local connections to be unusually long. Getting from the tram station to the Octagon with its 500 rental units takes 5 min on a bicycle, 15 min with an average wait time of 3.5 min on the Red Bus, and 22 min to walk. The obvious solution is a local bike share system, analog to the very successful Citi Bike system in Manhattan (see Figs. 3.26, 3.33 and 3.34).

Fig. 3.33 Rendering of a bike share station to be located at the tram station (Concept and Image courtesy: Roosevelt Island Studio at Hunter College)

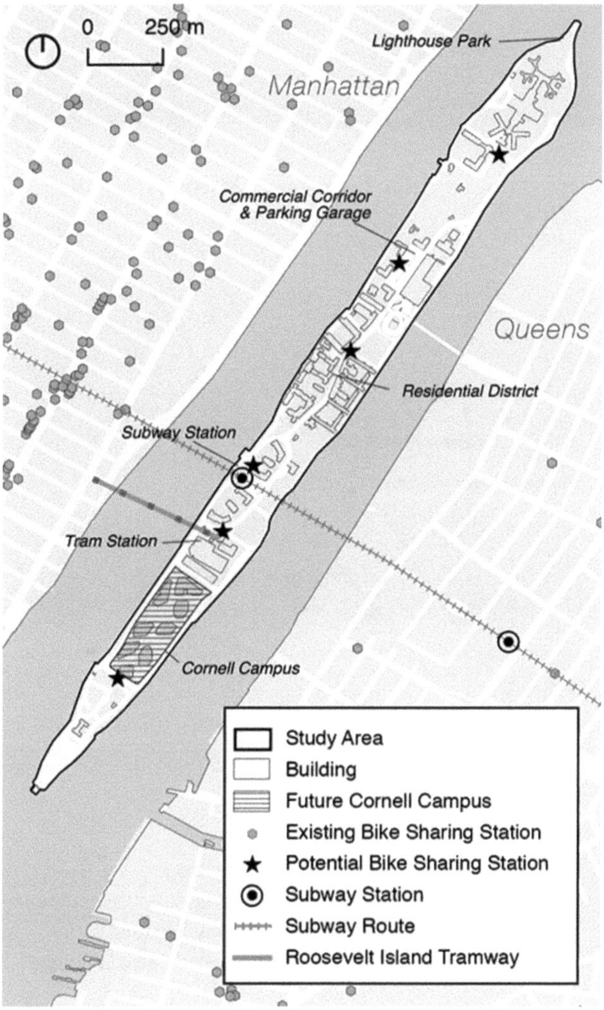

Fig. 3.34 Locations for suggested bike sharing stations

In spite of the extra subway stop on the island, and in light of the extra-long commutes on the island itself, residents were interested in a set of ferry services, each one providing one stop access to Manhattan (see Fig. 3.35). In a happy coincidence, this wish matches the Mayor's Office plans for increased ferry services throughout the city.

3.3.6.2 Cornell Campus

The City of New York effectively closed the Coler-Goldwater Hospital in 2011 and invited for bids to reuse the area for an international applied sciences campus. Cornell University, in collaboration with Tel Aviv Technion, won the competition to

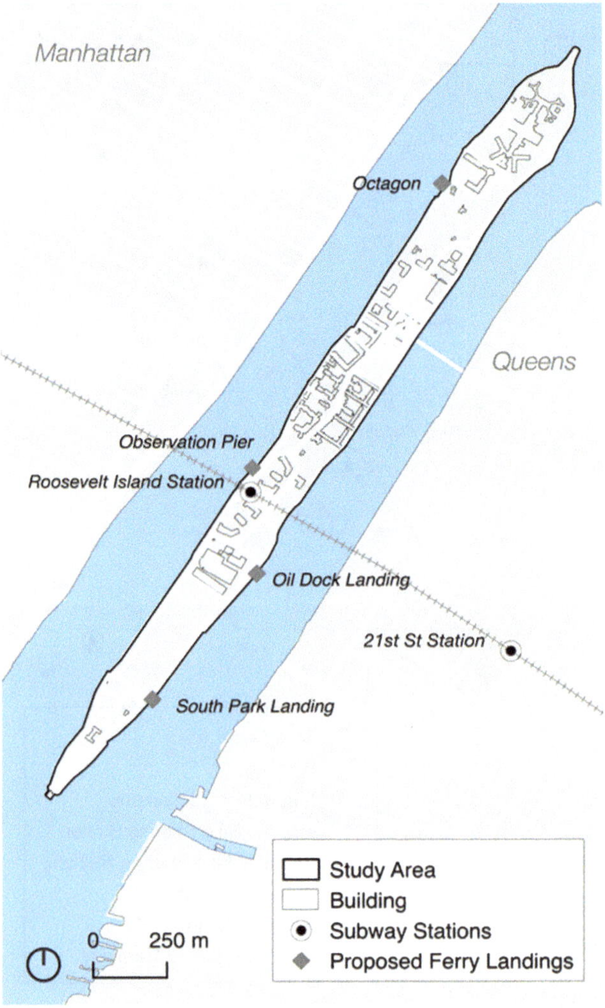

Fig. 3.35 Suggested locations for ferry landings on Roosevelt Island

build a non-biological applied sciences and engineering campus for graduate-level education that serves approximately 800 faculty and students and some 550 workers mainly in colocated businesses and a hotel.

The southern half of the project area is reserved for further development in the 2020s or 2030s and in the interim is planned as an open space with a nursery and a meadow (see Fig. 3.20). If fully developed, the figures in Table 3.3 would more than double. In an attempt to fulfill anticipated sustainability demands, the plans include heavy use of photovoltaic panels and geothermal wells. Although the academic component of the build-out is expected to be energy-neutral, an additional local gas plant is supposed to serve the other campus units (Fig. 3.36).

Table 3.3 First phase of campus build-out

Use	Square footage	Units	Stories
Academic	200,000		4
Residential housing			31
Faculty housing		104	
Student housing		338	
Residential total	300,000	442	
Businesses	100,000		5
Executive education center	170,000	225	13
Utility plant	20,000		
Parking		250	
Total	790,000		

Fig. 3.36 Phase 1 of the new Cornell campus

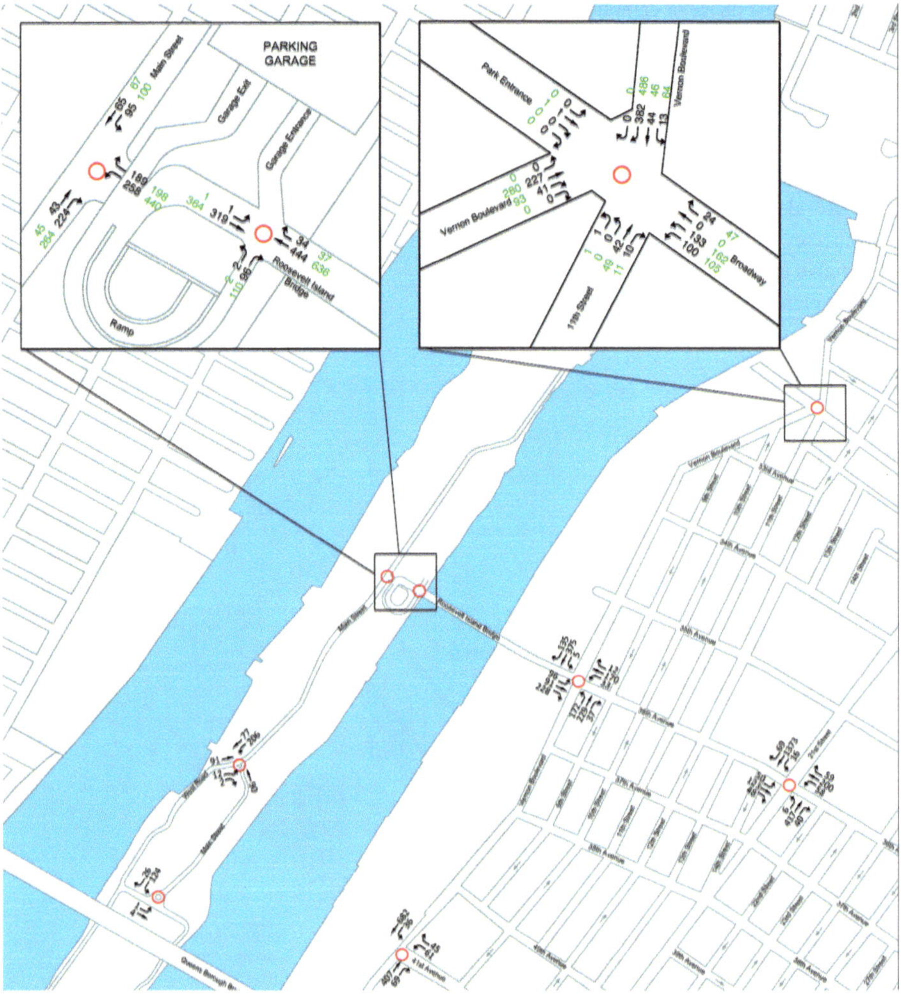

Fig. 3.37 Increase of vehicular traffic off the island caused by the new Cornell campus

One of the biggest impacts of the new campus would not as much be on the Island itself but in neighboring Queens, where the combination of local densification and the added demand by workers and students exceeds the capacity of the existing road network (see Fig. 3.37).

3.4 Concluding Comments

We believe that every plan, big or small, is as unique as is the place where it is applied to. We spent so much space on the description of our two case studies because it helps us to put the methods that we are re-familiarizing the reader with in the following two chapters into context. One of the major drawbacks of academic learning is just that, that it is academic. A good graduate program will teach students to learn techniques using real-world (messy) data, including collecting such data from scratch. But by necessity of the teaching environment, each method is going to be taught in isolation.

Once a planner works in an agency or as a consultant, the situation is suddenly not as pedagogically clear anymore and she will have to deal with conflicting interests at all levels. In a first step, she will have to get a lay of the land, similar to the introduction to our case studies but now on her own. Chapter 4 introduces a set of methods with occasional references to the case studies we just visited. Using either one of these case studies, the reader is invited to ponder which of the "planning grand" techniques would be applicable for Hunts Point or Roosevelt Island. Chapter 5 (while the two chapters can be read independently, they complement each other and should be read as one) will then delve deep into exploring the planning task-based on data-intensive methods. As we have shown here, it is very hard to get to know a place without some reliance on data, and it is virtually impossible to plan for the future without playing through some scenarios. While this is not a traditional methods book, we believe that every planner ought to have an overview of the methods discussed in the following chapters. At the same time, every tool in the planner's tool chest has to fit the purpose of the task at hand and be appropriate to the situational context as exemplified by our case studies here.

References

CSS (2008) Community Service Society 2008. Mapping Poverty in New York City. Pinpointing the impact of poverty, community by community. New York, NY

NYC 311 (2017) 311 Service requests. New York City Open Data. Online resource, https://nycopendata.socrata.com/Social-Services/311-Service-Requests-from-2010-to-Present/erm2-nwe9. Last accessed 5/7/2017

NY State (2015) Opiod hospitalizations. Statewide Planning and research Cooperative System (SPARCS), Albany. Online resource, https://www.health.ny.gov/statistics/sparcs/. Last accessed 5/7/2017

PlaNYC (2011). PlaNYC: A Greener, Greater New York. New York City: The Mayor's Office

US Census (2016). ZIP Code Business Statistics, Part of 2015 Business Patterns, Table CB1500CZ11. Available at American FactFinder (Census Bureau), http://factfinder.census.gov

Chapter 4
Planning Grand

4.1 Projecting into the Future

Regional-scale research methods concerned with long-range forecasting typically have time horizons of 25–50 years. This chapter covers a range of qualitative and quantitative research methods that can be classified as either descriptive or prescriptive approaches. Descriptive or extrapolative methods seek to describe objectively what the future will be or could be. Prescriptive methods adopt a more normative approach and are concerned with how the future should be. Pesonen et al. (2000) identify six categories:

1. Extrapolative methods such as trend analysis, time series, regression, econometrics, and simulation modeling are based on the notion that the future represents a logical extension of the past.
2. Exploratory methods typically use qualitative methods to structure possible futures, e.g., morphological analysis, relevance trees, mind mapping, and futures wheels.
3. Modeling approaches seek to describe the future by identifying the determining mechanisms of past events and how these influence the future. Examples include analogy analysis, technological sequence analysis, stakeholder analysis, and structural analysis.
4. Scenarios start with the basic premise that the future is unpredictable and, as a result, modeling will not lead to one future but to many different futures, each of which can be described in the form of a scenario.
5. Participatory methods such as cross-impact analysis, focus groups, and the Delphi technique rely on expert and stakeholder opinions and their insights about the future.
6. Normative methods investigate what the future should be and what steps are necessary to get to that future (backcasting).

L. Ramasubramanian, J. Albrecht, *Essential Methods for Planning Practitioners*, The Urban Book Series, https://doi.org/10.1007/978-3-319-68041-5_4

These six categories are by no means exclusive and a number of researchers advocate a multi-method approach when exploring the future.

4.2 Delphi

The Delphi technique was developed at the RAND Corporation in the 1950s and 1960s to improve the forecasts of a group of experts (Murnighan 1981). Although improved on in subsequent years and now in existence in many variations, the original Delphi technique is based on the three characteristics, *anonymity*, *iteration*, and *feedback*. The meeting of the experts is virtual, and the conversation is moderated. In an initial round, the experts are asked to identify the main criteria and give an estimate as to when they would be met. The moderator compiles these, anonymizing the responses and then shares them with the group. In a second round, participants are then asked to provide arguments for their forecasts. Here, it is important that the participants can see how their original opinion fits within the range of group opinions, which may alter their perspective and influence the arguments they make. Participants are asked to be as specific as possible so that moderators can develop a table of quantitative measures. In any (small) number of iterations, moderators then provide not just that table but also individual measures of deviation from the group consensus. The goal is to arrive at a group consensus within a relatively small amount of time.

The Delphi technique per se says nothing about the expert selection process. Experience tells us that the number of iterations should be kept to a minimum as even experts get easily bored by the process. The feedback is important but usually leads to the reigning in of outside opinions rather than to majorities changing their view based on the weight of the argument. As such, this method typically arrives at the statistical median of the original opinion – a fact that can be used to shortcut the whole procedure. Another benefit of the Delphi technique is that it is usually very hard to reach consensus among a group of experts. The anonymity and moderated feedback have proven to be a very efficient tool for reigning in people, who are not used be second guessed.

The Delphi technique is a nice mix of quantitative and qualitative aspects. The questions of the first round are typically on a nominal or rank scale, such as what the main drivers of future developments are or how desirable a particular development might be. As participants are asked to substantiate their claims, the method moves into the quantitative realm that is necessary to provide the statistical feedback. The moderators play an important role; not only do they anonymize the answers and compile the tabular feedback, but they must also rephrase individual contributions in such a way that they are unambiguous to the other participants. Experience in survey design, normalizing the topics so that there are no overlapping categories and, of course, the selection of appropriate experts can be quite challenging.

A relatively recent review of the Delphi technique, including a case study example, can be found in Gordon's (2009) chapter for the Futures Research Methodology compendium.

4.3 Futures Wheel

The futures wheel is a graphical decision-making tool that is aimed at identifying all conceivable impacts. It is particularly useful in the brainstorming stage of impact analysis but also serves for risk analysis on a qualitative level and could be categorized as a particular form of *mind mapping*. It was developed by Glenn (1972) and has been adopted by corporate planners and public policymakers to identify potential problems and opportunities, new markets, products, and services and to assess alternative tactics and strategies (Fig. 4.1).

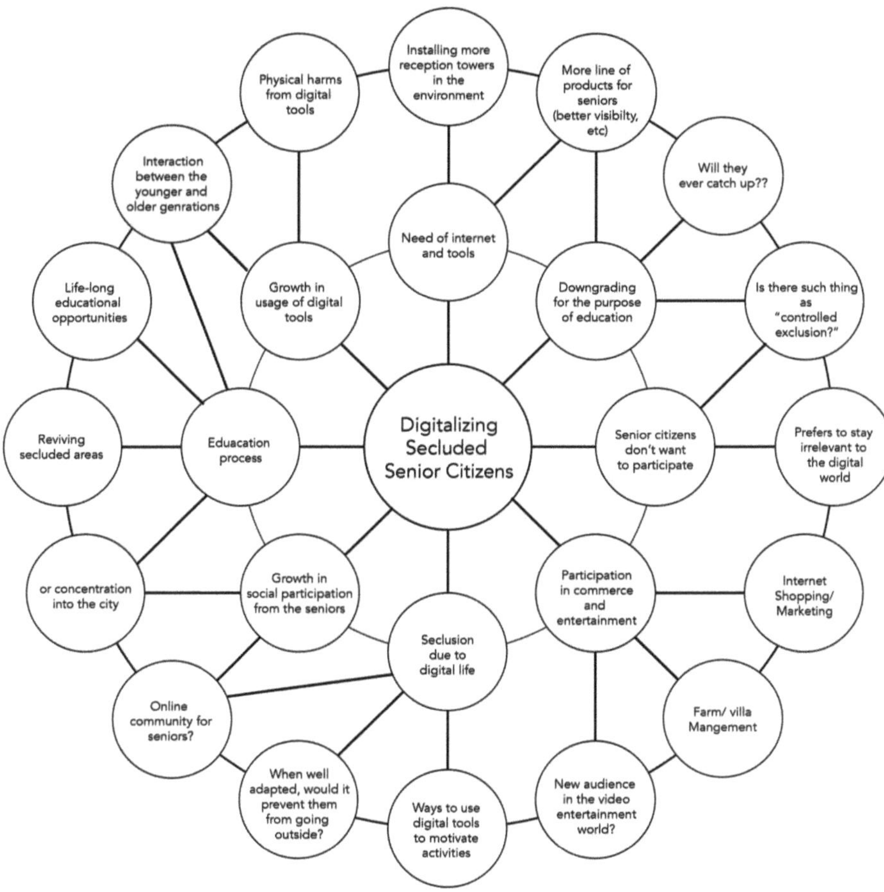

Fig. 4.1 An example of a futures wheel as part of a brainstorming exercise about how to increase digital access of senior citizens in secluded areas (From: Hwang 2016)

4.4 Scenario Planning

Scenario planning is a methodology employed to examine plausible, divergent futures based on uncertainty about drivers of change. This examination allows individuals and organizations to develop their capability and capacity to make robust decisions. Scenario planning is grounded in a qualitative approach that stands in contrast to quantitative forecasting tools that consider predicted futures. The key strength of the methodology lies in the discovery of potential or possible futures, including how decisions today could play out in years ahead. The ultimate goal of scenario planning should, therefore, be to build flexibility into decision-making.

Scenario studies have two principal purposes (van Notten 2003): exploration and pre-policy research. Exploration scenarios are primarily concerned with "learning, awareness-raising, the stimulation of creative thinking and investigating the interaction of societal processes" (van Notten 2003, p. 5). Meanwhile, pre-policy research "may propose concrete options for strategic decision-making and it is common in pre-policy research scenario exercises to offer implicit policy recommendations" (van Notten 2003, p. 5).

Pre-policy scenarios involve an extensive dialog with key stakeholders that become the focus of strategic responses. Scenario narratives can be used to identify common trends or themes that are consistent across different scenarios and so enable an organization to prepare its future by building flexibility into decision-making. Backcasting is a method often used in these types of scenario exercises to identify how the trends or themes play out over time to get to where they are in the scenarios. Backcasting can help organizations identify signals or "seeds" that indicate progression toward a scenario from the present, or it can help to identify a desirable position to aspire to and the steps needed to get there.

The World Economic Forum's (2014) Scenarios for Mongolia is an excellent example for such scenario planning, providing three different directions for the country given its mineral wealth and geographic position between Russia and China. The World Economic Forum used the scenarios to develop "common policy options" that would be robust under all scenarios and policy options specific to each of the scenarios. The scenario exercise stimulated strategic discussions among the country's leaders and was designed to be a continuing point of discussion into the future. In this way, the World Economic Forum's scenario exercise had the dual purpose of stimulating strategic thinking and discussion while at the same time identifying common responses to the scenario pathways.

Bishop, Hines, and Collins (2007) provide an excellent overview of eight categories of scenario-building techniques: judgment, baseline/expected, elaboration, event sequences, backcasting, dimensions of uncertainty, and cross-impact analysis and modeling. The latter two categories are encapsulated in Wilson's matrix (Fig. 4.2). "High-impact/low-uncertainty" forces are the relative certainties for which planning must prepare. "High-impact/high-uncertainty" forces are the poten-

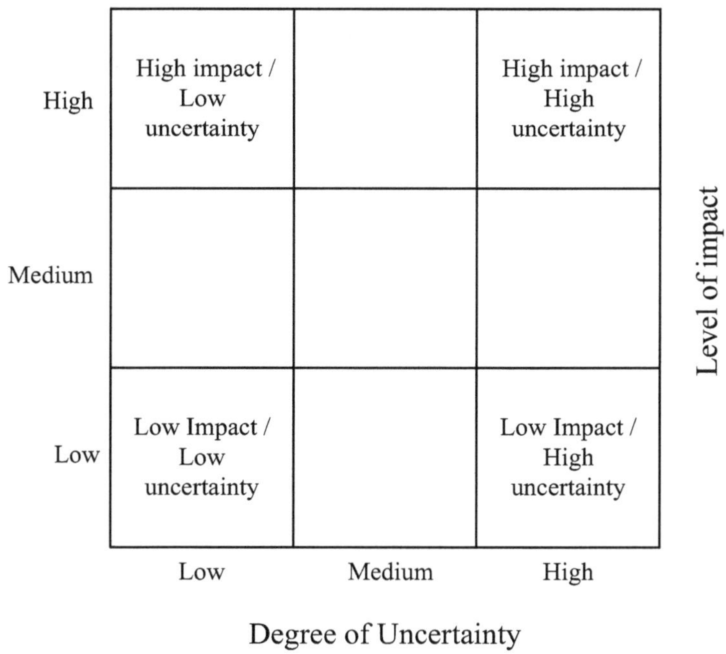

Fig. 4.2 Wilson's matrix

tial shapers of different futures (scenarios) for which planning should prepare. A state-of-the-art implementation of scenario planning is the program *CommunityViz.*[1]

4.5 Forecasting

A particularly important component of scenario building is forecasting, the modeling category of Bishop et al. above. The term is used in urban planning slightly differently from traditional science or engineering disciplines where the goal is to make predictions based on a trend. Forecasting in urban planning is more postmodern in that it involves the development of different models such as demand models, social and/or environmental impact analyses, as well as cost-benefit analyses of each scenario (Waddell et al. 2007).

A famous example is the four-step model in transportation planning, first developed in the 1950s for the metropolitan areas of Chicago and Detroit (Mitchell and Rapkin 1954). Based on land-use forecasting models of their own, the four steps in this urban transportation planning procedure each are made up of their own models for:

[1] http://communityviz.city-explained.com/

- Trip generation
- Trip distribution
- Mode choice
- Route assignment

Although still heavily used in metropolitan planning agencies, the four-step model is increasingly replaced by more individual-based activity models that lead naturally to the next category of simulation games.

4.6 Simulations and Gaming

An almost infamous example for the blurring of the lines between entertainment, education, and what-if scenario playing is the SimCity[2]™ series. Simulations are based on models that represent the key characteristics or behaviors/functions of the selected physical or abstract system or process. The model represents the system itself, whereas the simulation represents the operation of the system over time. By changing variables in the simulation, predictions may be made about the behavior of the system. It is a tool to virtually investigate the behavior of the system under study. A classic example would be a traffic simulation that helps to discern how behavior will change according to the set of initial parameters assumed for the environment.

Urban planning simulations are often based on agent-based modeling environments with explicit representations for land use and transportation. UrbanSim[3] and LEAM[4] are examples of large-scale urban simulation models that are used by metropolitan planning agencies for land-use and transportation planning. Academic journals like *Simulation & Gaming,*[5] published since 1970, and a set of publications by the National Academy of Sciences (CMSG 2010; Honey and Hilton 2011) illustrate how simulations and games can be used as an outreach tool (see Chapter 6).

4.7 Dealing with Bias

All of the above may sound very scientific and, hence, ring true. However, it is important that the planner, as well as participating citizens, is aware of the potential bias that can creep in at every stage. A planner, especially one with lots of experience, has through her personal history acquired a position on every subject matter. Experts, invited to participate in a Delphi study, are even supposed to have their

[2] http://www.simcity.com/.

[3] https://github.com/UDST/urbansim.

[4] http://www.leam.illinois.edu/leam.

[5] http://sag.sagepub.com/.

individual biases; a problem arises when they do not represent the full range of possible expert opinions. Communities have their own agendas, especially in the case of NIMBY issues.

Bias can also be hidden in the very method that is supposed to avoid all of the above examples of bias. Web surveys, for example, are prone to self-selection and hence a skew. Poor survey design may lead to unacceptable non-response rates either for the whole survey or for a subset of the questions (Pearson et al. 2010).

In his groundbreaking article on behavioral economics, Nobel laureate Kahnemann (1994) proved the inane cognitive bias in all forms of human forecasting. His and related research prompted the American Planning Association in 2005 to endorse *reference forecasting* to reduce inaccuracy and bias in forecasts. The method requires taking an "outside view" on the project being forecasted by examining similar projects, creating a distribution of outcomes for the reference class, and then positioning the project within that distribution (APA 2005).

This outside perspective, i.e., the inclusion of other similar projects in one's estimate of likely outcomes, also curbs what psychologists call the planning fallacy and optimism bias. In the grip of the planning fallacy, planners and project promoters make decisions based on delusional optimism rather than on a rational weighting of gains, losses, and probabilities. They overestimate benefits and underestimate costs, and thus planners and promoters pursue initiatives that are unlikely to come in on budget or on time, or to ever deliver the expected returns.

References

APA (2005) JAPA article calls on planners to help end inaccuracies in public project revenue forecasting. APA Press release on 7 April 2005, originally published at http://www.planning.org/newsreleases/2005/ftp040705.htm

Bishop P, Hines A, Collins T (2007) The current state of scenario development: an overview of techniques. Foresight 9(1):5–25

CMSG (Committee on Modeling, Simulation, and Games) (2010) The rise of games and high-performance computing for modeling and simulation. National Academies Press, Washington, DC

Glenn JC (1972) Futurizing teaching vs futures courses. Soc Sci Record 9(3)

Gordon T (2009) Chapter 4: The Delphi method. In: Glenn J, Gordon T (eds) Futures research methodology, vol 3. The Millennium Project, Washington, DC. Online resource, http://millennium-project.org/FRMv3_0/04-Delphi.pdf. Last accessed 1/2/16

Honey M, Hilton M (eds) (2011) Learning science through computer games and simulations. National Academies Press, Washington, DC

Hwang V (2016) Side-effect & side-shows: seniors and digitizing world. Vicky Hwang Lab Blog. Online resource, https://medium.com/vhlab/side-effects-side-shows-seniors-and-digitalizing-world-6834b2991e93. Last accessed 5/7/2017

Kahnemann D (1994) New challenges to the rationality assumption. J Inst Theor Econ 150(1):18–44

Mitchell R, Rapkin C (1954) Urban traffic: a function of land use. Columbia University Press, New York

Murnighan J (1981) Group decision making: what strategies should you use? Manag Rev 70(2):55–62

Pearson D, Hard E, Farnsworth S, Forrest T, Spillane D, Ojah M, Wojak K, Boxill S, Lewis C (2010) Improving accuracy in household and external travel surveys, Report 0–5711-1. Texas Transportation Institute, College Station
Pesonen H, Ekvall T, Fleischer G, Huppes G, Jahn C, Klos Z, Rebitzer G, Sonnemann G, Tintinelli A, Weidema B, Wenzel H (2000) Framework for scenario development in LCA. Int J Life Cycle Assess 5(1):21–30
Van Notten P (2003) Scenario development: a typology of approaches. Retrieved from http://www.oecd.org
Waddell P, Ulfarsson G, Franklin J, Lobb J (2007) Incorporating land use in metropolitan transportation planning. Transp Res A 41(5):382–410. https://doi.org/10.1016/j.tra.2006.09.008
World Economic Forum (2014) Scenarios for Mongolia. Retrieved from http://www.weforum.org/issues/scenariosmongolia

Chapter 5
Placemaking: Why Everything Is Local

5.1 Understanding Demographic Profiles

This section is about empirical work in support of the specific planning question. The increase in usefulness and specificity comes at a high price: each of the tasks described here is very labor-intensive; they cost a lot of time, and hence money and planners usually will have to scrutinize carefully whether the effort is warranted. On the other hand, it can be argued that plan making without any of the following is likely to be superficial and prone to limited acceptance by those we plan for.

We understand planning as an activity for and by people with "space" being the object rather than the subject of planning. If we want to understand whom the plans are for, then we need to know who the people are. Eventually, this will involve actually talking to and planning with the people we are trying to serve. And ideally, we know the area and their residents well enough to not have to go through preliminary surveys. The first step is often to look at the (recent) past as it is captured by either an official census or a large-scale public or private survey. Table 3.2 in Chapter 3 is an example for such a quick look at demographic trends.

In most countries, the main source for demographic data is the respective census department. Depending on the scale (of geographic extent and spatial resolution) of the area to be planned for, local authorities and/or private (both commercial and non-for-profit) organizations may have more appropriate or more up-to-date data. We use the term demographic in a rather expansive sense, i.e., any kind of data that helps us to understand, judge, and work for the populations we are trying to serve. We distinguish this from environmental data that describes the not person-specific conditions people live in/under (discussed in more depth in Sect. 5.3).

Demographic data then describes study populations as a whole and helps us to identify subgroups. At this stage, the goal is never to pinpoint individuals but generic characteristics of the populations we are dealing with. Basic and hence ubiquitously used examples include age, income levels and sources, educational attainment, ethnicity, and language spoken. Less common, not always available, but important for

L. Ramasubramanian, J. Albrecht, *Essential Methods for Planning Practitioners*,
The Urban Book Series, https://doi.org/10.1007/978-3-319-68041-5_5

certain types of planning are disabilities, religion, family status, and race. The latter is a hot-button issue, starting with those who correctly deny the physical (genetic) existence of race, moving to those who believe that racial differences are an issue of the past, and ending with proponents who also correctly observe that regardless of what racial classification we use, the conditions of certain groups of people with the same color of skin and the same set of discriminatory experiences have objectively not changed much over the past so many decades. This is not the place to discuss the issue of race in detail; however, we encourage the reader to peruse de Souza Briggs (2005), Sugrue (2005), and Gillette (2006) in the reference section of this chapter.

We would like to provide a special plug for the Integrated Public Use Microdata Series[1] (IPUMS) database at the Minnesota Population Center. This is the one-stop place to search for national and international demographic, health, and education data, in some cases going back for the last 200 years. We are using their National Historical GIS (NHGIS) on an almost daily basis (but the reader should browse their environmental or time use databases as well). While the US Census Bureau has already ceased to provide access to the decennial census of 2000, NHGIS not only has all the data but even aligns it to several previous generations of census geometries – thereby preventing major headaches for everyone who wants to study trends.

We are discussing demographic data here rather than in the previous chapter because the majority of planners work at the neighborhood level. Unfortunately, only a few variables are made available at the neighborhood or census tract level (with the exception of densely populated areas like our two case study areas in Chapter 3, census tracts usually are the size of neighborhoods). Most interesting variable combinations, as well as transportation or land-use-related data, are only made available at the coarser PUMA or county level. PUMAs or Public Use Microdata Areas are despite their name not really micro in nature, as they are required to have at least 100,000 people living in them (2015 average is 130,000) – far bigger than what we understand neighborhoods to be. The second caveat working with this kind of data is that it is usually quite old. The next two sections will deal with attempts to overcome the problem, but in general, when we work with (public) demographic data, we pay either on the spatial resolution side or on the temporal one. In other words, the more detailed we want our geography to be, the more averaged out over time will the data be. And many compilations, such as the extremely insightful Census Transportation Planning Package (CTTP) that we used for the determination of places of work for our Hunts Point and Roosevelt Island residents, are often 10 years old (AASHTO 2017). This is why planners have to increasingly resort to other Web-based data sources. Here, we need to distinguish (a) by price and arguably quality and (b) by accessibility. All the government and research-oriented websites offer data for free. In most instances, access to private/company websites is free only for sample data sets or a limited time – an attempt to lure the potential customer into a purchase. Often, the "free data" is formatted in such a way that it takes considerable effort to clean it for further use. Increasingly, rather than actually downloading all the data, data providers offer

[1] https://www.ipums.org/.

Table 5.1 Data sources

Organization	URL	Free	API	Comments
US Census Bureau	census.gov	Y	Y	
Federal agency consortium	webferret.com	Y	N	
University of Minnesota	nhgis.org	Y	N	NSF-funded long-term project
Investigative Reporters	census.ire.org	Y	Y	Other data for sale
Advameg Inc.	city-data.com	Y	N	Difficult to machine access
Datarealm	zipwho.com	Y	N	Difficult to machine access
The Nielsen Company	claritas.com	N	N	Market analysis specialist
ESRI	esri.com/bao	N	Y	See also their *community tapestry*

access via an application programmer interface (API). This sounds more complicated than it is; the skills to create customized access to such data from within a GIS can be acquired in a day and every planner should learn how to do that. See our book website,[2] for examples (Table 5.1).

5.2 Crowdsourcing

Crowdsourcing is an alternative to data compiled by others. It has the potential to be more detailed and more up-to-date than the data sources we discussed in the previous section. The higher degree of detail is achieved through the inclusion of local knowledge. In addition to the filter of life experiences, such local knowledge incorporates specific circumstances, events, and relationships that are not available through official data collection. In that sense, crowds can be experts, albeit more of locality than of academic subject matter. An example of such official data augmentation is the AirCasting project[3] in Hunts Point. A whole issue of the Urban Science journal was devoted to examples of crowdsourcing urban data (Wentz et al. 2018).

One advantage of crowdsourcing is that the data is extremely current. As such, planners may want to use crowdsourced data to extend the lifespan of the more traditional data sources discussed in Sect. 5.1. With that, we are now able to create a trend that takes us from the past to the present. The methods described in Chapter 4 then take us into the future. Another advantage is that crowdsourcing broadens the realm from which ideas can be collected. New Web-based forms of outreach allow for the inclusion of ideas that would never have been presented at a face-to-face public hearing. Virtual meetings that are disconnected in space and time empower voices to contribute (Starhawk 2011).

So what does a planner need to know about crowdsourcing, i.e., how is it done? The prime example for a well-established crowdsourcing platforms that every

[2] http://allthingsplanning.org/.

[3] http://brie.hunter.cuny.edu/hpe/2014/12/22/study-aims-to-determine-whats-in-the-air/.

planner should know about is OpenStreetMap[4] or, short, OSM. Former debates about the accuracy of that data source have been addressed by a myriad of published studies, and increasingly, official agencies from Berlin to New York City have adopted it for their own work and are sharing their latest updates with OSM (Haklay et al. 2014; ArrivingInBerlin 2017).

A nice example for how OSM data can, in turn, be used to provide the base for client/citizen suggestions is the "Suggest a location" page[5] of the bike sharing program Divvy in Chicago. A page like this one can now be cobbled together in an hour with readily available open source or enterprise tools (such as ArcGIS Online). A planner interested in serving this same community might then also avail herself of the publicly available data from Strava (2017). Originally a manufacturer of exercise equipment, the company compiles heatmaps[6] and provides an application programmers interface (API) to access data on aggregated pedestrian and bicyclists' travel patterns.

An example for a most basic yet core planning application is the Detroit-based but now nationwide operating platform makeloveland.com.[7] This kind of parcel-level data collection can realistically only be achieved by incorporating the public. In this case, the benefit for the local contributors are immediate, but crowdsourcing can also be successfully deployed to get free labor as for a good deed as the New York Public Library shows with their Emigrant City,[8] Building Inspector,[9] oral history,[10] and Map Warper[11] projects. Planners, who would like to get started truly at point zero, will find introductory tutorials on the geographical open data kit website.[12] Applications like Map Your World,[13] where children and young adults are collecting data show that this is not rocket science. A complete open source planning application, suitable for anyone with basic programming skills, is LocalData.[14] If the reader is overwhelmed by anything that looks like (programming) code, then she should look at Code for America[15] as a temporary source for help in setting up crowdsourcing and digital community engagement services.

A nice (hypothetical) example for how to crowdsource public participation in planning can be found in the second half of a 2009 article by Brabham. The idea of

[4] http://www.openstreetmap.org/.

[5] http://suggest.divvybikes.com/page/about.

[6] http://labs.strava.com/heatmap.

[7] http://makeloveland.com/.

[8] http://emigrantcity.nypl.org/.

[9] http://buildinginspector.nypl.org/.

[10] http://oralhistory.nypl.org/.

[11] http://maps.nypl.org/.

[12] http://geoodk.com/.

[13] https://s3-ap-southeast-1.amazonaws.com/myw-media/myw/wp-content/uploads/2014/10/Map-Your-World-Guidebook.pdf.

[14] https://github.com/LocalData.

[15] http://digitalcharlotte.org/code-for-america-lands-in-charlotte-the-goal-build-citizen-engagement/.

crowdsourcing in planning-relevant applications has been popularized in 2007 by Goodchild coining the term "volunteered geographic information" or VGI. The article was entitled *Citizens as sensors: the world of volunteered geography*, which leads us to our next topic, sensor networks.

5.3 Sensor Networks

There is plenty of planning-relevant information that cannot be directly observed; traffic counts being a prime example. Goodchild's notion of citizens as sensors alludes to the fact that a lot of data, especially in our age of *Big Data*, is collected automatically. Groups of autonomous measuring devices that are distributed over a geographic area are called sensor networks.

The majority of sensor networks have been deployed either by natural scientists (e.g., weather stations) or civil engineers, particularly in the transportation sector. From a regional planning perspective, the latter is of particular interest as they allow for real-time updates on traffic flow patterns, which in turn can be used to manage the signal system in what is then referred to as an intelligent vehicle management system.

Intelligence is a term that is quite often conjured in this context; the other term is *smart*. Based on the concept of the Internet of Things (Ashton 2009), all kinds of devices are now deployed to measure air pollution, noise, light, vibration, traffic, and meteorological data to alert planning agencies to local problems that affect urban activities and public health. Chicago's waggle[16] or Cambridge (MA)'s CitySense program are examples for the deployment of fairly universal sensors that can be adapted for all kinds of urban planning applications. Although based on Open Geospatial Consortium and ISO standards, these implementations are still experimental, and the market is very immature. Numerous large companies (IBM, Google, Siemens) are trying to establish themselves in this area, but in contrast to applications in the environmental science and homeland security realm, there are no off-the-shelf implementations for Smart Cities yet.

5.4 Understanding What People Do by Observing Their Actions and Behaviors

This section encourages planners to use the skills and techniques developed and used outside of mainstream planning – drawing from the disciplines of architecture, urban design, and environmental psychology. Some planners may dismiss these approaches as "soft" or "without rigor". Nothing could be further from the truth.

---

[16] http://wa8.gl/.

5.4.1 Behavior Maps

Architects and planners build spaces with a particular purpose in mind and are often surprised about the way they are then used. Ittleson et al. (1970) introduced us to the idea to literally follow the pathways of each individual entering an area of interest while taking note of what everybody is doing, how long they pause in front of or near what structure, and, not the least, how much interaction there is between each of the individuals observed.

The emphasis here is on *observation*; there is no interference with the individuals, no asking them why they are doing whatever they are doing. In a way, since we cannot presume to understand their behavior, the emphasis of behavior maps is more on the space and its elements. It is like observing the behavior of animated objects on a computer screen and trying to discern the rules according to which those objects move: speed, direction, number of turns, collision avoidance, as well as length of presence in the observed space, clustering/gathering, and interaction with elements of that space. Mere observations are sometimes sufficient to determine what environmental conditions act as connections and which ones serve as separators.

Behavior maps have been often used to help optimize the use of special environments such as children playgrounds, nursing homes, museums, or transit hubs. While the recording is still quite labor-intensive (for attempts to automate, see the following section), the analysis of such data has become a lot easier with the introduction of GIS in general and "space syntax" methods (Hillier and Hanson 1984; Jiang and Claramunt 2002) in particular.

A special subset of behavioral maps is places and traces mapping (Zeisel 1984). Any use of space leaves traces – be it footsteps, worn-out paths, discarded trash, etc. A place that shows no sign of use signifies that it is not used, which in itself is significant.

Where the local environment does not meet the needs of their inhabitants, people adapt them. This may be in the form of moving (street) furniture or by changing their intended use (stairs become sitting places). Places that are frequented by the same people tend to change over the time of day. Whyte's 1980 *Social Life of Small Urban Spaces* has not lost its relevance.

5.4.2 Sensors/Trackers

Increasingly, behavior maps can be created semiautomatically by handing subjects tracking devices that log people's locations at regular time intervals. Each one of these devices records to what amounts to a space-time path as depicted in Fig. 5.1 (Hägerstrand 1970; Kwan 2004; Neutens et al. 2012). Multiple tracks can then be overlaid to determine opportunities for interaction or to get a sense of the use (crowdedness) of space at different times. The passive counterpart to such tracking

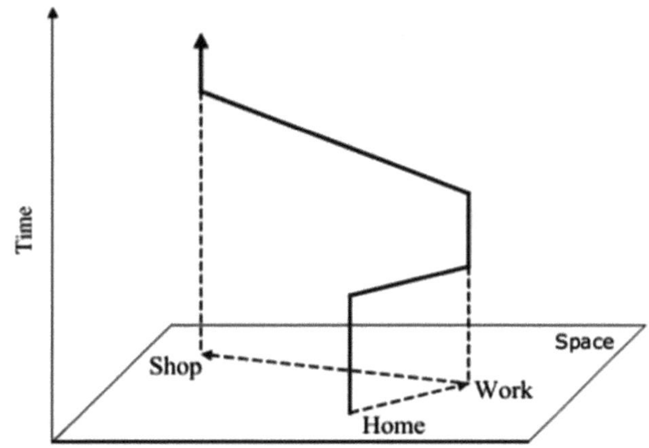

Fig. 5.1 Space-time path

devices is static sensors that just record the number and speed of objects/people passing. These are commonly used in all kinds of traffic counts, from street intersections to passengers at subway entrances.

5.4.3 *Participant Observation*

An extension of the behavioral techniques described above is to shadow consenting citizens during their daily routine. In addition to recording space and time, this very intensive method allows us to ask the participant why they are doing what they are doing. Only at this level (and the next), do we learn about the motivations and the reasoning process that causes people to use their environment in a particular way. For example, if we see someone waiting for ½ hour at a bus stop in winter, we might conclude that this person has no car; but the effect that this has on this person's other daily spatiotemporal decisions cannot be discerned unless we follow them throughout the day and learn about their decision-making in context. In addition to the time spent in the field (in order to assure "representativeness," a multitude of participants need to be observed on different days of the week and preferably in different seasons), transliteration, coding, and analysis of these observations require significant resources.

5.4.4 *Ethnographic Research*

The more time the observer spends in the field, the more her work approaches the realm of ethnographic research. The goal here is to experience life in the community by recording minutely all observations of everyone the researcher is encountering while living in the community. In anthropology, such studies are measured in years trying to immerse oneself to be part of the community. This is not realistic for planning purposes but can be approximated (a) by short-term multiple participant observation and (b) by encouraging members of the community to become active researchers. Participatory action research emphasizes this latter option, where the research/planning question is indigenous to the community and the traditional separations between the outside researcher (planner) and the community she serves are erased (e.g., Whyte, 1991)

5.5 Understanding People by Asking Them

The methods in Sect. 5.4 aim to minimize the effort on the subjects of our work. They are passive in the sense that we try to interfere as little as possible. The following set of methods, on the other hand, asks participants to reflect on how they interact with their environment. A point in case is perceptual mapping, where there is no objective space as defined by an outside observer.

5.5.1 Perceptual Mapping

Closely related to behavioral mapping, here there no aquarium within which we observe the movement of objects. Instead, we ask residents, workers, or tourists to create a map from scratch. It is up to the participant to select *what* they want to map and what the spatial relationships (distance, topology) among their objects are *according to them*. Perceptual maps provide enormous insight because it is those perceptions that form the basis for people's behavior; as such, they provide explanatory power that the behavioral maps lack. The downside is that perceptual maps are often as hard to interpret as dreams. It is very difficult to ascertain the meaning of what is drawn on such maps, and it requires a lot of interaction with their creators to assemble a standardized set of real-world features and relationships that can be translated into plan items. If there is a budget for it, then multiple perceptual maps can be overlaid in GIS (see Sect. 5.7) and the results quantified – which adds a valuable dimension to traditional official maps. A good example for such perceptual mapping is given in Sect. 2.6 of our Hunts Point case study.

5.5.2 Key Informant Interviews

Key informants are (local) experts, thought leaders, or anybody else who has firsthand knowledge about the community. They are not necessarily representative of a larger group of people but are in a position that allows them to reflect on the thoughts and actions of a particular group of people. Examples for key informants are local politicians, religious leaders, and people who are hired as experts such as transportation planners, ecologists, etc.

Key informants are a great resource at the beginning of a research project because (especially if one has the opportunity to interview a varied set of them) they are a great way to stake out the scope of issues at hand. A well-chosen set of key informants will provide you with the history, the main players, and all important aspects of your research. They do not replace your own understanding of the issue and lack all aspects of a quantitative study, but they work well as bookends, helping to cast one's own research agenda as well as to provide critical reflection at the end of the study.

5.5.3 Annotated Online Maps

There is a myriad of online maps out there that allow citizens to create their own sketches or just place markers on a map, which are accompanied by comment boxes or annotations. Typically, these are project-/task-specific, i.e., they are created with a particular research question in mind, be it to collect complaints, to solicit community input on a proposed development, or to facilitate online collaboration. US-based examples include Community Remarks®,[17] GeoCommons,[18] and Mappler.[19]

As in the above two subsections, these maps are qualitative in nature. Aligning them with traditional GIS data (Sect. 5.7), e.g., counting the number of submissions for a particular feature on a Web map, requires some efforts as there is some fuzziness to the specificity of the input. The biggest advantage of annotated online maps is their expressiveness and sheer attraction of such maps (if they are well-designed). They have the potential to be a great tool for spatial collaboration.

5.5.4 Surveys

Dedicated surveys are the most powerful tool to learn about people, their perceptions, and subsequent behaviors. Of the methods discussed in this section, they are the most quantitative ones, which are both their boon and their bane. The boon is that once we have enough survey responses, a well-designed survey can be analyzed with an arsenal of quantitative tools that fills many textbooks. The bane lies in the two qualifiers of the previous sentence. Designing a survey to really and unambiguously address one's research question is a difficult task. There are a number of good

[17] http://www.communityremarks.com/.

[18] http://geocommons.com/search.html.

[19] http://www.mappler.net/.

textbooks (Rea and Parker 2005; Gaber and Gaber 2007; Babie 2012) that try to teach the art of survey design, but in the end, it takes some practice.

The second caveat is the need to have a representative sample. For very simple surveys, one might get away with a hundred or so respondents. But more complex surveys often have to be balanced by gender, age, income, or location (e.g., x number of respondents per ZIP code area) and that combinatorial explosion leads to survey requirements in the thousands – a very expensive endeavor.

A good survey can cost hundreds of thousands of dollars or years of research time (often both). New online tools may be able to circumvent some of these obstacles if one manages to couch the survey questions in such a way that the online form "goes viral," i.e., respondents help to distribute the survey in a snowball system. Offline communities such as the elderly or visually impaired will, of course, remain outside of the surveyed populations.

5.6 Understanding Local Environmental Conditions

Planning is simultaneously about people and place. The latter is more than just a locational reference as described at the beginning of this chapter. The whole geography of a place matters, i.e., the climate, the culture, the tax code, design considerations, land-use codes, etc. All of these form the "environment"; in other words, this term encompasses all the physical, social, and cultural characteristics associated with a place – it is what makes a place unique.[20]

In a planning context, the most common environmental conditions are the housing stock; availability of services such as transportation, sanitation, education, health providers, groceries, etc.; characteristics such as crime rates, employment opportunities, and neighborhood amenities; as well as what is usually considered "environmental": green spaces, noise, air, or water pollution. Each of these categories is a stand-in for a multitude of variables. Housing stock, for instance, which is owned or rented, has an age, vacancy rate, maintenance level, etc. Even the building material is important, for example, when considering fire resistance or slope and aspect of the roof when calculating solar efficiency.

Static data is typically easier to get hold of and to process than data about phenomena that change over time. Transportation accessibility, for instance, is very much a function of time. Numbered bus routes often take different routes at different times of day and may not be available when a middle school student tries to get home.

Once the data is in place (and aligned to the same geography of reporting units), it is fairly straightforward to employ descriptive statistics and to look for correlations, say between environmental conditions and social or behavioral outcomes.

[20] Although, arguably, some cookie-cut suburban neighborhoods may lack uniqueness.

5.7 GIS

There are more GIS textbooks out there than on any other method used by planners. And it is only fair to assume that the reader has some experience with some GIS software. Our goal is not to introduce the reader to fancy GIS methods but to illustrate how GIS can be used in concert with all the other methods discussed here.

A first step is to situate GIS among all the methods of Chapters 4 and 5. We are discussing it here because from a planning perspective, GIS is still not very usable at the regional level. Writing from our own experience in Metro New York, handling GIS for a 20 county, 30 million people region requires enormous resources. So large that the City only started its own city-wide GIS operations only after the turn of this century, and there is still no agency that has a comprehensive GIS for the region. This does have very practical consequences: planning at the regional scale is by definition sector-specific. In other words, comprehensive use of GIS is eminently local.

Second, and this may initially sound like a contradiction of the previous point, GIS is essentially quantitative in nature – with all the advantages and disadvantages that this entails. As one of the authors of this volume tells their students, "computers are incredible fast – and incredibly dumb." GIS takes everything literally. If we want to create a buffer of 500 feet, then this is what the GIS does, excluding features that are 501 feet away, although for all practical purposes, the two distances are the same. Similarly, cleaning data so that different sources truly match (e.g., spelling of street addresses or the data type of census IDs) takes both a lot of discipline and the patience of a saint. Ideally, the reader is working with a GIS specialist and does not have to do all of this herself. But even then, it pays to know the pitfalls and to understand the time it takes even when working with those incredibly fast computers.

5.7.1 *Determining Data Needs*

One of the ironies of working with urban and regional planners is that they seem to be pretty bad at planning their projects. Before anyone sits down in front of a GIS, it is opportune to (a) develop a conceptual model of the task at hand, (b) define the data needs based on that conceptual model, and then (c) pare those needs down to what is available/affordable. The result of (c) is what we refer to as an implementation model, i.e., the ideal world of the conceptual model cut down to what is achievable given local constraints. This process is important to guarantee transparency, to create awareness of the limitations of a project due to lack of data or funds.

5.7.2 *Organizing Spatial Data*

Given the many dimensions of demographic data, it comes as no surprise that each of these comes as a file or a table. Planners need rudimentary database skills to handle that kind of data, to link from one table to another, to create indices, to transform data from text to machine-readable to table format, to create and understand metadata, and to deal with different locational references. All of these functions are part of geographic information systems (GIS) or more specifically geo-databases such as SpatiaLite[21] or PostGIS.[22]

The first thing that every planner needs to pay attention to is the need for the data to have a locational reference. Without a (or multiple) variable that links to the study area or a part of it, the demographic data is not usable. The locational reference can be an identifier that is spelled out in another data set; it may be a name that can be uniquely identified in a gazetteer, an address, or a coordinate.

All four locational references in Table 5.2 refer to the same real-world feature, but they actually link to different representations. The identifier refers to a US Census tabulation block *area* with well-defined boundaries. The name refers typically to a building but could also mean the property grounds that this building is located in. The address turns out to be a so-called vanity address. The mail will be delivered (albeit not to the abovementioned building), but there is no house number at the entrance of the building which happens to be located along a *line* referred to as Pennsylvania Avenue. Finally, the coordinate is in reference to a particular geodetic datum (an equation describing the shape of the earth) and geographic coordinate system describing a *point* at the intersection of the two main axes of the building.

Complicating the storage of locational references (and hence their lookup and use) is that the identifier and name are each usually stored in a single field, while the address is part of a multitude of fields describing country, state, city, delivery zone, street name, and house number, and coordinates are stored in two or three fields and typically require additional information stored in another file that specifies the geographic coordinate system used. Because so much mail is addressed to the White

Table 5.2 Locational references

Reference type	Example
Identifier	110010062021039
Name	The White House
Address	1600 Pennsylvania Ave
Coordinate	38.8976962,-77.0364835

[21] http://www.gaia-gis.it/gaia-sins/.

[22] http://postgis.net/.

House, it has its own ZIP code (20500), which in this particular case is different from the ZIP code area it is located in (20006).

Because there are so many different ways to describe locations and because, historically, the specification of locations was the responsibility of one authority, while the description of what can be found at each location is distributed, the geometric component of demographic data is often stored separately from the so-called attribute data. The specific link between georeferenced geometry and descriptive attributes (such as demographics) is known as the georelational principle. Based on the more generic idea from information science, it allows to link many different tables to one and the same geometric representation of the geographic feature of interest. With the proliferation of different (often private) data sources, this separation is however slowly receding, and we increasingly find data, where all components of a feature are stored in a single record. Figure 5.2 is an example for a single long string that describes the White House in Washington, DC. Here, the geometry is a fairly detailed polygon (area) with a bunch of attributes that the authors have made up.

This trivial example does not capture the demographic complexities of neighborhoods and cities. The US Census Bureau, for instance, lists under the header of people-based data: age, ancestry, disability, commuting to work, education, employment, family/relationship, health insurance, income and earnings, language, origins, poverty, race and ethnicity, and veterans. For each of these, there are dozens of tables and often permutations across these categories. To complicate matters, in order to prevent analysts from identifying individuals, those permutations (e.g., (1) married (2) taxi cab driver (3) from Somalia (4) with three children) are only available for larger geographic areas. As a result, it behooves planners to develop a data(base) organization schema that matches the planning question on hand.

5.8 Spatial Analysis

The term spatial analysis means different things to different people. Technically, it involves the use of statistical methods when working with georeferenced data. GIS is not the tool of choice to do that – although this functionality is slowly added to a number of mainstream packages.

```
{"type": "Feature",
  "properties": {"name": "The White House", "owner": "US
    Government", "occupant": "Barack Obama", "start_date":
    "2009-01-20", "end_date": "2017-01-20","dog": "Bo"},
  "geometry": {"type": "Polygon", "coordinates": [[-77.039336,
    38.895041], [-77.039336, 38.900051], [-77.033704,
    38.900051], [-77.033704, 38.895041], [-77.039336,
    38.895041]]}}
```

Fig. 5.2 Sample code for a geographic feature

More colloquially, the terms "spatial" and "geographic" are used interchangeably, and spatial analysis is the result of using analytical GIS functionality rather than the much larger set of data management functions. We distinguish analytical functionality by the kind of data that it is applied to.

Points	Techniques used to analyze an undifferentiated set of points, e.g., point pattern analysis
Spatial objects with attributes	Techniques that analyze an attribute matrix and reduce space to a square matrix of spatial relationships between pairs of objects, e.g., measures of adjacency or proximity
Networks of links and nodes	A range of techniques for analyzing networks in transportation and hydrology, based on attributes of network links and nodes
Spatial interaction models	Models of the interaction between pairs of objects, based on an analysis of the characteristics of origin objects, destination objects, and the spatial separation between them
Raster techniques	Methods of analysis based on the representation of continuous layers as rasters of cells and supported by the so-called raster GISs

5.8.1 *Vector GIS*

Of the data models above, point analysis falls squarely into the statistical realm. The same is true for most spatial interaction models, although there are a few that are more deterministic in nature – but their discussion would be way beyond the scope of this book. Spatial objects with attributes are our traditional vector model and the set of analytical tools here is as limited as it is ubiquitous: we are either dealing with some form of distance measurement or a small set of topological operations, the (in) famous buffers and overlays.

The reason for their ubiquity is (a) the familiarity of the vector model to anyone who has ever looked at a map and (b) the very fact that the set of functions is limited. The strength of vector-based analysis lies in the fact this small set of operations can be concatenated into workflow models that allow for intricate spatial filters and subsequently spatial decision support systems.

5.8.2 *Network GIS*

Networks of nodes and links form the basis for what is often confused with vector GIS but in fact is a very different kind of GIS. Networks can indeed be formed from the points and lines of a vector GIS, but the data organization is different, and more important, the set of analytical functions is very different. The graphs that network GIS is based on do not know of areas or polygons. There are no property lines, no parcels, no land use, all mainstay of a planner. The easiest way to conceptualize the difference is to think of a network GIS as if it were some form of subway map.

The nodes and links form graphs that may or may not be connected, as for example, when we are trying to depict two different utility networks. Yes, we want our houses (represented as nodes rather than areas) to be connected to the networks, but we do not want the gas and electricity networks to share nodes.

As in the vector model, analytical operations are limited to distance and topological measures. Given the difference in the data model though, the implementation is very different and in most GIS rather crude. The best implementations of network functionality are found in dedicated software packages that sometimes may call themselves GIS and offer some limited GIS functionality but really are graph-based tools such as SNAP,[23] the very expensive Palantir,[24] or Pajek.[25] The most appropriate dedicated package for applied planners is TransCAD.[26] Robin Lovelace's sustainable transport planning with R package (STPLANR 2017) is as of 2017 the most promising non-commercial solution.

5.8.3 Raster GIS

The one form of traditional GIS that most planners stay away from is raster GIS, which is ironic because from an analysis perspective, the raster data model offers by far the widest range of analysis functions. To see why, it is helpful to recall the notion that computers are very fast and very dumb. The raster data structure is extremely simple; all spatial relationships are implicit and based on the specification of the number of rows and columns as well as the size that each raster cell represents in geographic space. This basic spreadsheet-like matrix structure allows for very fast computation. Coordinate-based vector GIS, on the other hand, requires complex geometry calculations that not only bog down the computer but are also algorithmically really hard to implement (which helps explain, why the set of analytical operations in vector GIS is limited).

Raster GIS is often associated with image processing and natural science applications, which is correct but too limiting. The raster model is particularly useful for planning when we want to translate vector data into surfaces or when we want to apply fuzzy overlays. A common example of surfaces are cost distance calculations, but they are also applicable when we want to identify catchment areas (e.g., around schools or hospitals) or if we want to create more realistic representations of population distributions by applying dasymetric mapping techniques (as was done in Fig. 3.25). All of these are just examples for the advantage of changing the data model to better fit one's conceptual model. The real advantage of working with raster data is the kind of modeling that we can perform with it.

[23] http://snap.stanford.edu/snap/.

[24] https://www.palantir.com/.

[25] http://mrvar.fdv.uni-lj.si/pajek/.

[26] http://www.caliper.com/tcovu.htm.

One of the main drawbacks of vector GIS is that it is very good at representing a cadaster-like snapshot of a given geography, but it is not suitable to represent change. If we want to represent (rather than just visualize in the form of an animation) spatially differentiated change, then raster GIS is the tool to use. One of the big advantages of not having to deal with geometries is that we can now apply any kind of modeling equation vertically to cells across multiple layers or horizontally to neighboring cells. Anything that can be expressed mathematically can be calculated with raster GIS, and the same holds for multivariate statistical analyses. We can apply all kinds of distance calculations, including traditional network analysis, by using auxiliary layers as lookup table for what constitutes a node or a link.

A major reason for switching to the raster model is the ability to perform weighted overlays on as many layers as one could ask for. This makes complex spatial decision support systems and Geodesign (see next section) much more straightforward to implement than in vector GIS. Finally, the raster format used in GSI is the same as for cellular automata, a modeling technique widely used in land change science.

5.8.4 Visualization Versus Analysis

One common misconception is that software, which is capable of displaying phenomena in three dimensions or in form of animations, is also capable of analyzing in 3D or across time. For off-the-shelf software, this is only marginally true. There is limited volumetric analysis in some vector GIS, but with the exception of very expensive GIS applications in mining and defense, there are no GIS applications on the market that use true three dimensions or have built-in analytical routines for routing across three dimensions or performing spatiotemporal overlays. Useful, as such functionality would be, its implementation is still the realm of PhD theses rather than a planner's desktop.

5.9 Geodesign

No modern planning methods book would be complete without a section on Geodesign (there are multiple spelling versions, in one or two words, CamelCase, etc.). It combines tools from computer-aided design (CAD), landscape design (also one of the disciplines that GIS originates from), GIS, and modeling. Similar to the variations in spelling, there are also many different interpretations; there are a myriad of interpretations as to what constitutes Geodesign. Acknowledging its origins in McHarg's and Steinitz' systems thinking approaches of the 1970s, it is probably easiest described by a list of necessary and desirable characteristics.

5.9.1 3D

Geodesign relies on visual impact and has a focus on looking as "real" as possible. To do this, it borrows from CAD and landscape design and adds the notion of fly-throughs. It is hard to discuss this characteristic without the next one because they are so tightly integrated. Given the discussion of 3D GIS, true three-dimensionality is certainly more on the wish than requirement list, but the motivation to go 3D is certainly there.

5.9.2 Visualization

This single header deserves attributes such as "fancy," photo-realistic, etc. At a minimum, a Geodesign project ought to provide visualizations from many interactively chosen perspectives and at multiple scales. The end goal is to mimic Hollywood-style interactive videos that allow the participant to experience different design solutions.

5.9.3 Simulation

One of the purposes of Geodesign is to provide the client with scenarios that are based on calculations with real-world data. The Geodesign workshop[27] held at the University of Washington in 2015 is a nice example, as it worked with King County data on a number of scenarios that workshop participants had to work through.

5.9.4 Participation

Built into the design philosophy is the workshop character described above. Affected community members are supposed to participate in all steps of the design process, and by playing with scenarios and experiencing the visual outcomes, both learn and eventually decide on the optimal implementation strategy.

5.9.5 Geography

This one is not as obvious as it seems, given the CAD and landscape design sources, where the connection to a real-world setting is not always a prerogative. Geodesign, however, requires a GIS input, if not setting. Geography is seen as context-setting

[27] http://depts.washington.edu/pgist/Geodesign2015.

and as a determinant. Here, the distinction between mere spatial (geometries) and geography (people and place) becomes important again. Participants are not just supposed to be wowed by the visualizations but to have researched the implications of each scenario on their own lives.

5.9.6 *Sustainability*

Environmental sustainability was not as much front and center in the early years of Geodesign but now plays an ever more important role. As such, Geodesign is developing from a method to a school of thought, an approach that brings together aspects from a number of disciplines. It is no coincidence that the first Geodesign master program in the United States (at Philadelphia University) stresses its post-postmodern philosophy. While the philosophical foundations are contested, the emphasis on issues like minimizing a project's carbon footprint is widely shared.

5.9.7 *Resources*

This one is not part of any official definition yet probably the most important ingredient. To accomplish even a minimal set of the six characteristics above takes an enormous amount of time, especially on the preparatory side. During a Geodesign workshop, it is essential that the scenarios can be run within an acceptable amount of time, which requires significant computing resources. Workshop planners may wish to run the whole workshop in the Cloud, e.g., Amazon Web services, Google Computer, or Microsoft Azure. There are, at the time of writing (early 2017), no software packages that combine all the necessary capabilities. The closest we have at this point is CommunityViz that we mentioned in Chapter 4. Setting up a Geodesign workshop will realistically take six person months, although each of the steps in 9.1 to 9.4 could be implemented individually in about one person month.

5.10 Planning in a Brave New World: Reflecting on Data Junk and Other Thoughts on Planning

The past few years of Big Data have changed the world of planners irrevocably. On one hand, there now is more data available to us than we hoped for only a few years ago. On the other hand, there is a deluge of messy data that confuses not just planners but also the concerned public. Open data laws have challenged public officials to struggle with ways to make data both intelligible and protect privacy. Releasing data in unintelligible form (non-editable formats, poor spatial resolution, noncompatible identifiers, etc.) was for a long time the bureaucrat's way to deal with such demands. But the public's expectations with respect to data access are constantly evolving and rising, and we now observe in most administrations a melee between different business units, some of which have adjusted to the new open world and others who have not.

The long-term solution will be well-managed public data infrastructures and the education of those custodians who learned their lesson that data is power. In the meantime, planners would do well to familiarize themselves with some data science techniques that are used to reformat and recast data to their needs. Familiarity with tools like Python and Jupyter[28] will not only make a planner's work a lot easier but also greatly enhance her career opportunities.

5.11 Experts Versus Non-experts

The discussion of methods in this chapter might have given some readers the idea that we have expert-driven advice on one side versus community-driven "feelings" on the other. Crowd-sourced data collection, for example, may be perceived as qualitatively inferior, although we already tried to nip this notion in the bud in our discussion of OSM. Let us, therefore, look at another argument, the wisdom of the masses.

Surowiecki (2005) has shown that when it comes to the accuracy of predictions, the average of independent predictions of non-experts tends to beat the predictions of experts. Similar to what we outlined for the Delphi technique, it is important that the data is collected independent of the bias of the next person but a planner would be well advised to listen to the community. Rather than treating expertise as a sliding scale, we see it as the summary of different experiences and approaches, like the metaphor of the blind men and the elephant.

5.12 Geographic Constraints

All of the data that goes into any of the analyses above has to have a spatial footprint, i.e., a locational reference that in most cases describes an area of interest and its subdivisions (see also Sect. 5.1 in this chapter). This spatial support trips up many beginners for two very different reasons. One is that the reporting units may not match, the other that the underlying geographic coordinate systems do not match.

[28] https://jupyter.org/.

5.12.1 Reporting Unit Mismatch

Different agencies have different reporting units that fit their respective mission. Examples are school districts, police precincts, ZIP code areas, or electoral districts. It would not be particularly challenging to list a dozen of these for a city like New York City. Attempts to reason across such dissimilar areas lead to the modifiable areal unit problem or MAUP (Openshaw 1984). Figure 5.3 exemplifies the reporting unit mismatch. Any statistical analysis of the two crime patterns depicted here will result in very different outcomes, depending on whether the base unit is a census tract, a voting district or a police precinct. If one has access to finer-grained data, e.g., at the address level, then this allows to reaggregate one's data to the desired target layer. Where such data is not available, one can try to use techniques such as dasymetric mapping (Mennis 2003) to, for example, redistribute populations, where it is safe to assume they do not exist (parks, water bodies, etc.). This was done in Fig. 3.25, where the census tract population of Roosevelt Island was redistributed to the building footprints using the fair assumption that households surveyed by the US Census Bureau have to live in a residential building and that the number of apartments in a building is a fair way to distribute the population across the buildings. Any such calculations are, however, based on assumptions that will have to be verified. The reader is advised to also check whether the original scales for which the data was compiled are roughly the same; as a general rule of thumb, any combination of spatial footprints will deteriorate the output to the coarsest resolution input.

5.12.2 Coordinate System Mismatch

All geographic data (as opposed to abstract civil engineering drawings) has to be specified in units of measurement (degrees, meters, feet, etc.) and given a coordinate system origin. Lack of a proper specification and, depending on what software

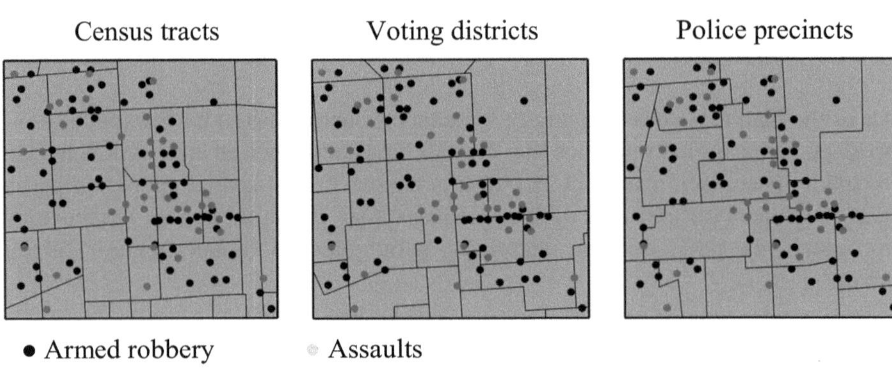

Fig. 5.3 Example for the modifiable area unit problem

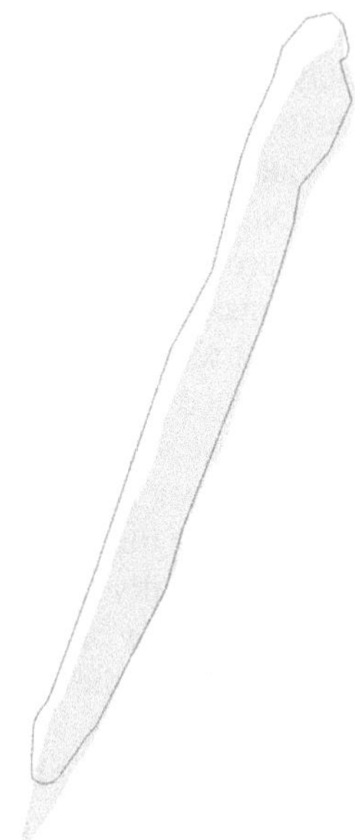

Fig. 5.4 Coordinate system mismatch

is used, translation into a common coordinate system will lead at best to errors but more likely to the data sets not to align at all. See Fig. 5.4 for the effect of a coordinate system mismatch in case of the now-familiar Roosevelt Island. Both data sets are from the US Census Bureau, one for 1990 Decennial Census data, the other for 2011–2015 American Community Survey data. The difference between the two versions of the southern tip of the island is a whopping 250 m.

5.12.3 Lying with Maps

Mark Monmonier wrote in 1991 a little book entitled *How to Lie with Maps* that concentrates on purposeful distortions of visual representations of geospatial data. Assuming that our readers are acting ethically, the bigger challenge rests with creating visualizations that are not misleading and appropriate for intended audiences. Guidelines, including do's and don'ts are well documented in Krygier and Wood's Making Maps (2011).

5.12.4 Limitations of a Data-Driven Approach

This book is written from a quantitative-friendly perspective. Yet, the authors are well aware of the limitations of an overreliance on data, especially in the age of "Big Data" (see previous section). Similar to the real estate mantra *location, location,*

location, all the approaches discussed in this chapter are dependent on context. If anything, then all the data collection and manipulation is intended to serve as context-setting – a shortcut allowing the planner to get a feel for the situation that she is about to interfere with.

One obvious limitation of the non-experiential approach is the lack of history that provides attachment to a place, no matter how dysfunctional it may look from the outside. Only highly specialized surveys and techniques discussed in Sects. 5.2 and 5.3 will accomplish that. Related to that is the fact that GIS methods are essentially good for taking snapshots of a given situation. GIS is very cumbersome (at best) if we want to represent processes or change. At a minimum, any attempt to do that requires very careful planning on the data management side.

Another limitation of current data-driven approaches is the fact that the data we have is usually not the data we need to represent our conceptual models, i.e., our understanding of the phenomenon we trying to study. Thus, there is a mismatch between what the data can actually tell us and what we want to believe it can tell us.

Finally, the use of geospatial data, especially of vector data, assumes that the numbers (whether they are coordinates, distances, or attribute values) are exact. This is, however, almost never true. Accuracy-appropriate GIS analysis would require fuzzy reasoning techniques that are way too cumbersome for traditional planning applications. The results of GIS analyses should hence be taken with a grain of salt.

5.13 Concluding Comments

This chapter is chock-full of pointers to a plethora of methods that a planner ought to be familiar with. Let the reader be reminded that ours is not a methods book nor do we advocate for planners to become bogged down with all the data massaging techniques that we are mentioning here. Every method has its place and the choice is dictated by the planning situation. This is why we introduced two case studies in Chapter 3 and have been constantly referring to them both here and in Chapter 4.

The reader is also reminded of the fact that several of the techniques introduced in these two chapters were not used in either case study. Partly, this is due to the fact that a typical graduate planning studio or a real-world planning project does not have the resources to conduct the more intensive methods like scenario building or Geodesign. We will, in the following two chapters, continue to drive home the notion of context. Every planning task has a budget and a situational context that goes beyond mere geography. Planning methods have to match the needs of the community that we are planning with (Chapter 6) and have a whole range of implementation constraints that influence the choice of methods discussed in the past two chapters.

In a way, we, whose bread and butter job it is to teach quantitative and qualitative methods, are striving to write an "un-methods" book. Sure, every planner should be familiar with what we presented here (and if she is not, the following section provides plenty of pointers for further reading), but at the same time, it is important not to lose sight of the real purpose of planning. Technologies can be a smokescreen, and some planners might even want to hide behind the "objectivity" and anonymity of the computer. The following two chapters will provide the counterweight and drive home the fact that methods are necessary means to an end but means that should be applied judiciously. Such judgment calls will become easier to make with experience, and Chapters 6 and 7 are intended to fast-track this learning process.

References

AASHTO (2017) Census transportation planning products. American Association of State Highway and Transportation Officials, Washington, DC. Online resource, http://ctpp.transportation.org/. Last accessed 5/7/2017

ArrivingInBerlin (2017) A map made by refugees. Berliner Stadtmission, Berlin. Online resource, https://arriving-in-berlin.de/. Last accessed 5/7/2017

Ashton K (2009) That 'internet of things' thing. RFID J. Online resource, http://www.rfidjournal.com/articles/view?4986. Last accessed 5/7/2017

Babie E (2012) The practice of social research, vol 13. Wadsworth, Belmont

Brabham D (2009) Crowdsourcing the public participation process in planning projects. Plan Theory 8(3):242–262. https://doi.org/10.1177/147.3095209.104824

de Souza Briggs X (ed) (2005) The geography of opportunity: race and housing choice in metropolitan America. Brookings Institution Press, Washington, DC

Gaber J, Gaber S (2007) Qualitative analysis for planning and policy: beyond the numbers. Planners Press American Planning Association, Chicago

Gillette H (2006) Between justice and beauty: race, planning, and the failure of urban policy in Washington, DC. University of Pennsylvania Press, Philadelphia

Goodchild M (2007) Citizens as sensors: the world of volunteered geography. GeoJournal 69(4):211–221. https://doi.org/10.1007/s10708-007-9111-y

Hägerstrand T (1970) What about people in regional science? Pap Reg Sci Assoc 24(1):6–21. https://doi.org/10.1007/BF01936872

Haklay M, Antoniou V, Basiouka S, Soden R, Mooney P (2014) Crowdsourced geographic information use in government. Report to GFDRR (World Bank), London

Hillier B, Hanson J (1984) The social logic of space. Cambridge University Press, Cambridge

Ittleson W, Rivlin L, Proshansky H (1970) The use of behavior maps in environmental psychology. In: Proshansky H, Ittelson W, Rivlin L (eds) Environmental psychology. Holt, Rinehart and Winston, New York

Jiang B, Claramunt C (2002) Integration of space syntax into GIS: new perspectives for urban morphology. Trans GIS 6(3):295–309

Krygier J, Wood D (2011) Making maps: a visual guide to map design for GIS, vol 2. Guilford Press, New York/London

Kwan M (2004) GIS methods in time-geographic research: geocomputation and geovisualization of human activity patterns. Geografiska Annaler: Ser B Hum Geogr 86(4):267–280. https://doi.org/10.1111/j.0435-3684.2004.00167.x

Mennis J (2003) Generating surface models of population using dasymetric mapping. Prof Geogr 55(1):31–42

Monmonnier M (1991) How to lie with maps. University of Chicago Press, Chicago

Neutens T, Delafontaine M, Scott D, De Maeyer P (2012) A GIS-based method to identify spatio-temporal gaps in public service delivery. Appl Geogr 32(2):253–264. https://doi.org/10.1016/j.apgeog.2011.05.006

Openshaw S (1984) The modifiable areal unit problem. CATMOG 38, University of East Anglia, Norwich, UK

Rea L, Parker R (2005) Designing and conducting survey research: a comprehensive guide, vol 3. Jossey Bass, San Francisco
Starhawk (2011) The empowerment manual. A guide for collaborative groups. New Society Publishers, Gabriola Island
STPLANR (2017) Sustainable Transport Planning. Functionality and data access tools for transport planning, including origin-destination analysis, route allocation and modelling travel patterns. R package maintained by Robin Lovelace. Internet resource: https://cran.r-project.org/web/packages/stplanr/index.html, last accessed 23 October, 2017
Strava (2017) Engineering with a global dataset. Online resource, http://labs.strava.com. Last accessed 5/7/2017
Sugrue T (2005) The origins of the urban crisis: race and inequality in postwar Detroit. Princeton University Press, Princeton
Surowiecki J (2005) The wisdom of crowds. Anchor Publishing, New York
Wentz E, Conrow L, Fischer H (eds) 2018. Crowdsourcing urban data. Urban Sci 2(1). Online resource, http://www.mdpi.com/journal/urbansci. Last accessed 5/7/2017
Whyte W (1980) The social life of small urban spaces. Conservation Foundation, Washington, DC
Whyte WF (1991) Participatory action research. Sage Publications, Thousand Oaks, CA
Zeisel J (1984) Inquiry by design: tools for environment-behavior research. Cambridge University Press, New York

Chapter 6
Civic Engagement

6.1 What Is Civic Engagement?

One of the overall goals of this book is to offer planning practitioners a curated list of research and analysis methods, techniques, and approaches to support and inform their day-to-day work. The methods discussed in this book are generally agnostic of specialization insofar that they can be used by land-use planners, transportation planners, urban designers, and so on, although each specialist may gain value from a subset of methods that are discussed in Chapters 3 and 4. This chapter focuses exclusively on the complexities of civic engagement. All planners must familiarize themselves with civic engagement practices.

Invoking a traditional dictionary definition, *civic*, is an adjective that is related to the activities or obligations of individuals in a public forum and/or concerning issues that relate in some way to public life, while *engagement* is a noun that typically conveys a legal or moral obligation to do something. In other words, a student who speaks at a college-wide governance forum is exercising her civic duty, so also is an individual who votes in a local election.

As we grow up and participate in everyday activities beginning with school, we learn quickly that we are part of a more complex societal fabric, outside of our immediate familial networks. We are linked by geographies, ideas, aspirations, and values; this recognition becomes the first step to understanding our obligations and responsibilities to care for and protect our shared ideals and interests. In democratic societies, the process of choosing our elected officials is often referenced as an obvious exposition of an individual's civic duty. In addition to the actual voting, civic actions also include canvassing for votes and participating in the public sphere to demonstrate in support of, or against causes or candidates. In this chapter, we view participation and engagement in electoral politics as an essential but rather narrow framing of civic engagement. Community-oriented civic engagement activities are not *apolitical* per se. Civic engagement is shaped by dominant sociocultural norms, providing individuals and groups the needed motivation to participate or to act.

L. Ramasubramanian, J. Albrecht, *Essential Methods for Planning Practitioners*,
The Urban Book Series, https://doi.org/10.1007/978-3-319-68041-5_6

However, activities carried out by the public that create and give voice to social and political concerns by questioning and challenging established social norms and practices are also an important and necessary aspect of civic engagement in contemporary America and the world.

The term "social capital" is often used to characterize and explain civic engagement, which became popular after Robert Putnam brought it to the forefront of the public's attention in the early 1990s. His arguments are organized and presented in the book *Bowling Alone* in which he speculated about the causes and consequences of the decline in civic engagement in America (Putnam 2000). Social capital refers to networks of association and trust, a kind of social glue that brings people together to work on volunteer activities, social causes, and creating a feeling of belonging, albeit for a short time. Much has been written about social capital, and these ideas and concepts previously found favor in global organizations like the World Bank (Bebbington et al. 2006).

Some of these social networks tend to be insular – in that the groups develop a sense of belonging and identity because they perceive themselves to be distinct from others around them. For example, immigrants to the United States often connect with immigrants using familiar associations of the regional geography/language/religion of their country of origin as a marker of trust. This "bonding capital" is useful to build and strengthen a group's identity and can help accomplish some civic objectives. However, many more civic engagement objectives can be achieved when groups with different interests connect with each other. This is not so hard to understand, once we think about our lived reality – we are all members of many groups because of our multiple identities, and we belong to different groups – practicing planners are also sports enthusiasts, and there can be a great diversity of sports and teams to choose from. Planners are also parents, caregivers, churchgoers, nature lovers, and runners. The list can go on. The kind of social capital that helps create associations between heterogeneous groups fosters civic engagement activities that are outward facing that can serve a larger social purpose.

In this chapter, we argue that all planners have a responsibility, an obligation, and the skills to support and nurture civic engagement, i.e., to create opportunities for the public to participate in civic activities. In other words, we encourage planners to recognize community organizing and mobilization outside of the sphere or electoral politics as civic engagement. This requires professional planners to inform, educate, and empower the public to actively participate in community decision-making at a variety of ways. This commitment is akin to a way of working, rather than a single task to be checked-off a list. Civic engagement and the role of planners have in facilitating that engagement is an ongoing process and part of the planners' own code of ethics. More about ethics later, in Chapter 7. For now, consider our claim that civic engagement can influence a project or program's success. Likewise, in the twenty-first century, it is a truism to state that good projects and ideas can sometimes be derailed because of poorly managed civic engagement processes.

The next section critiques the two established frameworks that have been used by planners to create a culture of civic engagement. Understanding the strengths and limitations of these frameworks is essential for all planners, especially for those planners who are at the start of the careers.

6.2 Origins of Modern Civic Engagement

Historically, some sections of the public have always participated in planning decisions – the upper echelons of American society were engaged with the City Beautiful movement as well as other subsequent reform-minded planning actions that began at the turn of the last century (Hall 1996). Elites used their time and resources to support "good" government-led planning and design. The middle class and the poor were not consulted; they were treated as ignorant or otherwise incapable of making meaningful decisions about the quality of life in their neighborhoods and cities.

Modernist ideals of progress demanded the dramatic reconfiguration of the built environment. Changes included the creation of impressive civic and public works and the development of robust transportation infrastructure and road networks that essentially cut through neighborhoods creating ruptures in the urban and social fabric. Modernist planning predicated the social, political, and economic transformation of cities and regions as determined by and dependent on transformations of the built environment.

The received wisdom of the time was that the harm done to a few neighborhoods and communities was balanced by the gains for the city and region. The people who were left out of these decision-making processes did not have an easy way to be heard. Although individuals with formal education expressed their dissenting opinions by writing opinion pieces and letters to the editor of major newspapers in much the same way we do today, it was a sad reality that once planners and politicians aligned together to accomplish noble goals – considering the interest of the majority and looking ahead 10, 15, and 20 years ahead – it became practically impossible to stop projects from moving forward.

Yet, everyday people were never fully complicit or compliant in the face of oppressive planning regimes. In big cities, tenants organized rent strikes to protest rising rents and used a combination of legal and public relations strategies to be heard. It took some time before the pattern of these displacements became all too apparent – that the interests of poor and working-class neighborhoods were being sacrificed.

The civil rights movement of the 1960s transformed and energized the voices of protesters, helping to launch the antiwar movement, the women's movement, and the gay rights movement, the environmental movement, and the disability rights movement. While the major cities Boston, New York, Chicago, and San Francisco all undertook "slum clearance" and massive redevelopment projects, the trend also affected smaller cities like Milwaukee, Portland, and New Orleans. Urban renewal projects came under critical scrutiny, and preservation of individual neighborhoods and communities became a way to organize and energize protest.

In her 1961 book, *The Death and Life of Great American Cities*, Jane Jacobs wrote a polemical but prescient book that critiqued the penchant for urban planners and architect to introduce a sense of order into city life – she used her experiences from Greenwich Village in Manhattan to reveal the social order and the security embedded in a bustling neighborhood that (at the time) was integrated across age and class lines if not by race. Lloyd Rodwin, a renowned MIT planning professor, reviewed her book[1] in the NY Times. In the review that appeared on November 5, 1961, he wrote:

> [Her book] fuses ineffectual elements of discontent into a program that can pack quite a wallop. It won't matter that like the reformers she criticizes, she has little sympathy for persons who want to live differently from the way she thinks they ought to live; nor will it matter that some of her own proposals (on the planning process, for example) come straight from the planners she criticizes; and that some of her cherished reforms, however tentatively advanced, are as romantic and "utopian" as those she rejects. The same holds for transparent gaps and blind spots, such as her blasé misunderstandings of theory and her amiable preference for evidence congenial to her thesis. In short, except to the miscellaneous victims and the academic purists, it won't matter that what this author has to say isn't always fair or right or "scientific."

Jane Jacobs used her voice as a citizen and engaged in a variety of actions to challenge and push back the power of planners. She is celebrated over a half a century later, because she succeeded! In 2017, a documentary *Citizen Jane: Battle for the City* celebrates Jacobs, the "non-expert" who wielded her power over and thwarted the grandiose aspirations of the master planner, Robert Moses. Matt Tyrnauer, the director, says that contemporary audiences in the United States and throughout the world can learn a lot from this epic battle. He says, "Jacobs tells us that we must be skeptical. We must look and listen for ourselves and then act to make the changes that will help our communities improve and thrive. You can't leave it to the 'experts.' There are no experts. The expert has to be you."[2] This is the planning challenge we described in Chapter 2. Tyrnauer appears to be saying that one's instincts and feelings about things can replace facts and objective analysis. He is not alone. We don't debate his view of the world here. However, complete adherence to Tyrnauer's reasoning will leave planners with limited options, privileging some publics, but undoubtedly marginalizing others.

Many books have been written about the David versus Goliath battle between Moses and Jacobs.[3] For our purposes, suffice it to say that the context of participation

[1] Rodwin, L. 1961. Review of *Jacobs, J. 1961. The death and life of great American cities*, 458 pp., New York: Random House, published in the New York Times, November 5th, 1961. Available from NYTimes Archives.

[2] Brynes, M. 2017. Why the Jane Jacobs vs. Robert Moses battle still matters, Atlantic City Lab, A Q&A with Matt Tyrnauer, director of Citizen Jane: Battle for the City. Available at: https://www.citylab.com/politics/2017/04/why-the-jane-jacobs-vs-robert-moses-battle-still-matters/523125/. Retrieved April 30, 2017.

[3] The Guardian, April 2016, Story of cities #32: Jane Jacobs v Robert Moses, battle of New York's urban titans https://www.theguardian.com/cities/2016/apr/28/story-cities-32-new-york-jane-jacobs-robert-moses. Retrieved April 15, 2017.

in the 1960s and 1970s was essentially reactive – attempts to stem the onslaught of a pro-development agenda that was supported by political liberals and conservatives alike at the time. Along with the activism of Jane Jacobs, the work of Paul Davidoff who helped to establish the concepts and principles of advocacy planning and Sherry Arnstein who delivered a blistering critique of established protocols for government-mandated public participation exemplifies the benefits and limits of reactive citizen participation.

6.2.1 Davidoff and the Advocacy Planning Model

Paul Davidoff (1930–1984), a planner and lawyer, felt strongly that the prevailing model of the rational-comprehensive model of planning was far from value-free. Rather than being "neutral" and representing the best interests of the public at large, he felt that the plans particularly large-scale planning projects put forward by city agencies tended to favor and benefit some while causing harm to others –the planning process was stacked against the interests of poor and marginalized communities, those who did not have a seat at the table.

In addition, Davidoff felt that it was impossible for an individual planner or a team of planning professionals to create plans that could clearly balance the interests of different groups, particularly those that held opposing viewpoints about a thorny planning issue. Davidoff argued that the differences and dichotomies of positions, especially value conflicts, would become more apparent if each planner or planning team advocated for the interests of one group or one issue – for example, the interests of renters or more broadly the interests of low-income people. Drawing upon his legal training, Davidoff further reasoned that these competing viewpoints, when argued by experts (planners), would allow better decisions to be made on the merits of the case. His arguments are summarized in his seminal article *Advocacy and Pluralism in Planning* (Davidoff 1965).

Several senior planning scholars have discussed Davidoff's contributions to the field,[4,5,6,7] and some of their views are summarized here. There are many benefits to undertaking advocacy planning. Advocacy planning helps planners clarify project goals, objectives, and intended outcomes. Good advocacy planning prevents

[4] Checkoway, B. Paul Davidoff and advocacy planning in retrospect. Symposium introduction. *Journal of the American Planning Association*. 60, 139–143, Apr. 15, 1994. ISSN: 01944363.

[5] Marris, P. Advocacy planning as a bridge between the professional and the political, part of a symposium on: Paul Davidoff and advocacy planning in retrospect. *Journal of the American Planning Association*. 60, 143–146, Apr. 15, 1994. ISSN: 01944363.

[6] Peattie, LR. Communities and interests in advocacy planning, part of a symposium on: Paul Davidoff and advocacy planning in retrospect. *Journal of the American Planning Association*. 60, 151–153, Apr. 15, 1994. ISSN: 01944363.

[7] Hayden, D. Who plans the U.S.A.? a comment on "Advocacy and pluralism in planning", part of a symposium on: Paul Davidoff and advocacy planning in retrospect. *Journal of the American Planning Association*. 60, 160–161, Apr. 15, 1994. ISSN: 01944363.

planners from making generic statements that are mere platitudes by inserting rigor in their analyses. Furthermore, planners are encouraged to take on a normative and activist stance toward addressing the needs of their neighborhoods and communities, rather than serving as mere functionaries who are content with implementing existing rules and regulations. Advocacy planning used/uses the adversarial approaches inherent in the legal system to address complex societal challenges such as racial segregation, urban renewal, and displacement. Yet, we note that the flaws inherent in this legalistic approach create its own set of new challenges.

In the last 50 years, advocacy planning has evolved and has become "professionalized." Advocacy planning now relies on outside experts, ultimately creating teams of expert planners who are "hired" to argue or champion different policy positions. As individuals interested in championing distinctive positions, these experts are less interested in resolving problems than they are about solidifying arguments and allies to support specific policies. Advocacy planning results in a wide range of intended and unintended outcomes. There is a great human cost to advocacy planning that affects individuals who may be at odds about the issue. There are additional costs to the neighborhood, community, and society: for instance, delayed projects negatively affect residential property values and force small business owners to go out of business while abandoned or boarded-up properties create unsafe and unsanitary conditions for people who live in the neighborhood. Collectively, the uncertainty associated with advocacy planning as it is practiced causes a great deal of anxiety to both sides – project proponents and opponents. Although it is not the intended outcome, when a "win" eventually arrives, it can feel like a "loss" for all other competing positions, creating permanent divides that are unsuitable for long-term community development.

6.2.2 Arnstein and the Ladder of Citizen Participation

Sherry Arnstein's 1969 article *A Ladder of Citizen Participation* is both a primer and critique on government-mandated efforts at citizen participation. Her ladder consists of eight rungs, moving from manipulation and therapy (nonparticipation) and further upward through informing, consultation, and placation (various degrees of tokenism), and culminates with partnership, delegated power, and citizen control (various degrees of citizen power). Arnstein's typology is a result of her concern about the fuzziness of the dominant terminologies associated with citizen participation – particularly "maximum feasible participation." The ladder she put forward allows us to unpack the concept, and she states explicitly that the typology is a purposeful simplification to "illustrate the point that so many have missed – that there are significant gradations of citizen participation."

Although Arnstein states that her typology can be generalized in different institutional and situational contexts, her primary understanding of the pros and cons of citizen participation came from her experiences with the Great Society programs of the Johnson administration, including the Model Cities program. In Arnstein's typology, the partnership rung of the ladder is characterized by genuine power sharing between community and government. This requires community residents to be organized and mobilized and to have the skills to understand the complexities of local management and governance and the resources to solicit expert technical assistance as and when needed. Delegated power allows for the community to hold decision-making authority or veto power over large sections of the plan through participation in boards or governing councils. Citizen control transfers the governance of local services or programs such as schools to the hands of citizens; many of these proposals were experimental efforts at direct democracy, bypassing preestablished frameworks of representative democracy such as elected city councils.

Arnstein's framing of the planning process as a dichotomy between governments as working with/for/against monolithic communities is a vestige of the past. Contemporary planners would do well to remember that Arnstein's citizen participation ladder is an elegant simplification. Blindly emphasizing the rungs on the Arnstein ladder limits an understanding of the various ways the present-day public engages in planning and decision-making.

6.2.3 *Discussion and Critique*

In the second decade of the twenty-first century, it is now 50 years since advocacy planning first hit the planning lexicon. It is over 40 years since Arnstein argued that government needed to be held accountable for its actions through active citizen participation.

Planners should celebrate the visible and systemic changes that have occurred since the 1960s because of the ideas put forth by Davidoff and Arnstein. The way planners are prepared has changed significantly; all planning schools now emphasize the value of public participation. In the United States, public participation is required by law in many, if not all projects, programs, and policies. Local government agencies routinely partner with a range of nongovernmental agencies to address complex issues such as affordable housing, social services provision, economic development, and neighborhood revitalization. Many states have also passed "sunshine laws" that limit planning decisions being made behind closed doors. While "manipulation" and "participation as therapy" persist, they are more routinely identified and vilified in the public sphere. The development and growth of digital technologies have expanded access to data and information, making it easier to hold elected officials and government agency employees accountable. From 2009 to 2016, the federal government expanded access to government data and information, in accordance with a federal commitment to transparency, participation, and collaboration.

We contend that the challenges that we face as a society now and in the next 50 years demand different strategies and tactics to reenergize planning processes. One of the limits of the work of Davidoff and Arnstein is that they focused very much on how to "fix" problems with the way things were being done at the time. Their approaches were reactive, responding to a particularly volatile time in American history, when trust in public institutions was low. Their approaches established an adversarial relationship between government planners and the people, a persistent challenge.

How should advocacy planning evolve? What lessons have been learned from the wins and losses attributed to advocacy planning over the years? When is it prudent to use the adversarial tactics of advocacy planning? Is citizen control (the highest rung of the Arnstein ladder) appropriate for all planning projects? Can providing information to the public be dismissed as mere tokenism? Is it fair that the city is held hostage by the delaying tactics employed by one interest group? These are the questions that contemporary planners should be asking, rather than relying exclusively on strategies and tactics that have worked in the past.

The concerns that were raised by Davidoff and Arnstein during the 1960s are not fully resolved, and as a society, we continue to struggle to address deep social injustices, environmental, health, and security challenges. Yet, it is crucial that planners propose and implement new ways of connecting with the public, acknowledging societal, cultural, and technological shifts that have influenced our everyday lives and will continue to do so.

6.3 Managing Change

A standard description of planning is the United States will most likely state:

> Good planning helps create communities that offer better choices for where and how people live. Planning helps communities to envision their future. It helps them find the right balance of new development and essential services, environmental protection, and innovative change[8]

Presently, public planning is a collaborative endeavor; it is focused on helping people manage change in a systematic and orderly way. Planners who are on the front lines, charged with empowering people from all walks of life, understand and come to terms with change and how it will impact their day-to-day lives.

Change can be daunting, even when it is desired. The creation of a plan is an integral component of change management. Although many planners work for and represent government agencies, planners also serve the nonprofit and private sector entities. In Chapter 2, we discussed three major global challenges – urbanization, demographics, and climate change, that create predictable physical and social consequences including the need to combat sprawl, create and maintain public infrastructure and transportation systems, plan for an aging society, plan for social and gender inclusion, address environmental quality, and design for climate resilience.

---

[8] American Planning Association, Colorado Chapter. http://www.apacolorado.org/what-is-a-planner. Retrieved April 10, 2017.

Planning students often hear that like politics, all planning is local. This is true to some extent, because the most intense debates about managing change occur at the local level. In this book, we urge planners to recognize and consider the global drivers of these local changes, for example, the planned closing of a local grocery store in a community may be a result of out-migration in the same way that need for new housing is a result of population growth in the same areas. Planners should address quality-of-life concerns such as solving the challenge of access to shopping when the only store has closed at the local level while simultaneously taking a systems approach to their work.

Some scholars question this commitment to quality-of-life planning. Radical academic scholars often exhort practitioners to struggle with the complexities of racialized policing and other social injustices that affect vulnerable populations in the public realm. New planners are often confused: they are labeled as the "enemy" by the left, simply because they signed up to be employed in their chosen profession, and at the same time pilloried by the right because they advocate government oversight and avoid mindless adherence to free market thinking. *Our sincere advice is to encourage planners to read, reflect, and learn from these critiques but continue to focus on problem-solving*. Practitioners must recognize that planning is about avoiding binaries of any kind – planners can address quality-of-life concerns and simultaneously strive to address serious social justice concerns through their work. One of the ways to do both is by practicing civic engagement strategies on and off the job.

Practicing planners should be the first to acknowledge that the feelings of anxiety and, in some instances, extreme trepidation that are experienced by the public are real, whether there is tangible data and evidence available to support these feelings or not. In this chapter, we explore how a planner especially a newly minted planner with limited access to her own networks and professional ties and one that feels vulnerable about her own job prospects and advancements can stand up to the established hierarchies.

We propose that by elevating the quality of the planning discourse, planning practitioners can help facilitate difficult but necessary conversations about change. Planning practitioners can reposition themselves as facilitators of dialogue between different interest groups, rather than representing one position over another. Planners become mediators and translators of information to help explain the differences between policy positions and over time can help to build a consensus about the scale and type of change that the community can bear.

We feel strongly that individuals in neighborhoods and communities cannot stay in a permanent state of opposition with each other. In small communities, it is practically impossible to navigate life in this way. Citizen activism and organizing must give way to decisions, projects, programs, and policies, and new institutional and governance structures are required to manage them. The strategies and tactics of reactive and hostile community mobilization are not suited for institution building which is better served through proactive civic engagement.

Consider climate change as an exemplar of a major planning issue that affects a wide range of communities in the United States and throughout the world. East-coast communities, whether they are year-round residential communities, heritage towns, working-class enclaves, or working waterfronts, will experience rain events, storms, and consequent impacts such as storm surges and flooding more frequently. Our observations and conversations in coastal communities tell us that long-term residents in these locations are aware of, and attuned to these changes. If we used the traditional advocacy planning model (applied in a stringent way), we would pit some groups within these communities against other groups. Railing against the government may not be a helpful approach either. In this context, a concerted effort is needed to address immediate concerns as well as plan for long-term solutions. Although different policies and approaches are needed for every coastal community, the principles of collaboration and consensus building are essential to resolve and manage the real impacts of climate change now and in the future. Since the government is a partner, rather than a leader in the planning effort, the participation should be managed in a way that is sustainable even after the government's formal obligations have concluded.

In Chapter 3, we discussed two case study examples, Roosevelt Island and Hunts Point, both neighborhoods in New York City. Having worked in these diverse neighborhoods in our teaching and research, we propose some civic engagement strategies that will facilitate consensus building and future-oriented planning.

6.4 A Framework for Twenty-First Century Civic Engagement

Creating a principled approach to participation wherein all members feel welcome and invited to participate is critical to the success of any project, program, or policy. Planners understand that they must engage with everyone, easier said than done. For planners in the beginning of their career, or even for those who have planned many interventions, we present a framework, informed by research and practice to create and sustain a progressive civic engagement agenda.

6.4.1 Understand the Rich Diversity of the Community

Be warned! This is not a trivial or easy task. Relying on publicly available data is a place to start, but nothing beats establishing a personal familiarity with the streets and alleys, the open spaces, the public facilities, and the various attributes of a community that you are planning for. We remind you to focus on understanding the community's composition along racial and class dimensions, without ignoring other demographic, cultural, and spiritual dimensions.

The data we have assembled about Hunts Point and Roosevelt Island should provide you with some guidance on how to develop your own community profile. For large cities like New York, a wealth of community profile data is readily available. Do not let the volume of data lull you into complacency. Data quality is elusive, because every piece of data is driven by a question that made its collection or creation possible. The framing of questions may have influenced the quality of the data set. We encourage field observations that include temporal and spatial variation. Get out of your car! Walk and bike, whenever possible so that you can experience the environment in different ways.

Understanding the community is essential to define the study area or the zone of intervention for a particular project, program, or policy. The conventional wisdom about defining a study area boundary may have to be renegotiated to have an adequate representation of all sectors of the community that may have a position about a proposed change. It is useful to draw a broader boundary than a narrower one so that a planner can be more inclusive than less inclusive.

There is an art to drawing study area boundaries – often, planners consider preexisting political boundary definitions which may or may not serve the purpose for a study. Physical geography sometimes can be a useful determinant of the edge, but again this is not always the case. People cross geographic boundaries like rivers and streams or human-made boundaries like freeways to access shopping or entertainment opportunities. It may also be useful to think about commuting via walking, biking, or driving in determining study area boundaries.

Since we are talking about pragmatic realities, we want to remind practitioners that a "paying client," be they a government agency or a private developer, may insist that a boundary definition be drawn in a prespecified way. In this situation, the planner should work very hard to understand the motivation behind the client's reasoning – we have found that in many instances, clients attempt to draw boundaries in a misguided attempt to avoid conflict. It is important that planners understand the motivation behind these "arbitrary" decisions so that the best advice and guidance can be offered. To the extent possible, it is the planner's responsibility to advise the client in favor of being inclusive, choosing to embrace potential conflict, rather than exclude it in the process of defining a study area. Ultimately, the planner and planning team have the freedom to gather data and conduct analyses outside the study area and include this information in their analyses to assess second-order effects.

6.4.2 *Promise to Engage the Whole Community and Keep Your Promise*

This is a simple and yet an elusive concept that appears to challenge many planning practitioners. Often, planning practitioners begin a project focusing on the need to engage the "difficult" groups – those that are expected to be in opposition to a proposed project. We unequivocally state that everyone who lives and/or works in a community should be consulted about a project or program that is being proposed. This decision has implications of how planners plan their outreach strategy. There are entire catalogues of public participation methods and techniques available to the planners interested in working with the public (e.g., Creighton 2005), and we encourage planners to familiarize themselves with them.

It is useful to make a public commitment to whole-community engagement because it creates an awareness that the project will approach engagement in a different way and that it is not a case of business as usual. It also allows the various actors engaged in the project remember that they have made a commitment to a more engaged, inclusive, and innovative approach to public involvement.

The public is a not a homogeneous or monolithic group. It is useful to consider planning with and for multiple publics. Rather than segment the population by race and/or ethnicity alone, it may be useful to segment in a way that cuts across these boundaries. Homeowners are an important segment but so, too, are renters. It is useful to include property owners in a conversation, but it is equally important to recognize the individuals who own properties may not be residents of your community. Sometimes, small business owners are left out of important conversations because they rent, rather than own their business locations. If the debate is about education and the expansion of a local school, it is still useful to engage adults without children or parents who do not send their children to that school. Our comments may fly against conventional wisdom and potentially increase the costs of doing community outreach (to be discussed later in this chapter), but we believe it is well worth the effort in terms of potential gains in creating and sustaining a cross section of the community that is interested in the future well-being of the place they have chosen to live, work, or visit on a regular basis. Once a planner clearly understands that there are multiple publics, she can design a community engagement strategy that is tailored to each of these groups.

6.4.3 Develop Civic Engagement Principles

In our view, community engagement must be approached in a principled and ethical way. One way to articulate a clarity of vision and approach is to lay out some principles of engagement. Examples of principles can include equity, respect, and inclusion. For example, if we believe in the principle of *equity*, then community engagement processes and protocols would pay special attention to fairness, ensuring that different segments of the public have the same (similar) opportunities to express their opinions. However, truly honoring the principle of equity would require the planners and the planning team to make additional efforts to ensure a diversity of views. The principle of *respect* is valuable so that people or groups holding seemingly unpopular views or positions can be heard. The principle of *inclusion* suggests that the process will not privilege nor disadvantage individuals or groups because of their geographic location or gender orientation or because of a policy position they hold.

Many other principles can be developed, depending on the scope and scale of the project. Two principles that practicing planners may want to consider are the principle of relevance and the principle of proficiency. The principle of *relevance* suggests that the advice and suggestions provided by the people of a community should be carefully focused to meet the scope, context, and needs of the project. In other words, the focus on relevance allows for individuals who may disagree on one subject (say, e.g., electoral politics) to agree on a shared vision about the need for additional investments for the central business district. This is a challenging principle to uphold in public forums and town hall meetings. The principle of *proficiency* allows the planning team to solicit, support, and use the skills, knowledge, and expertise that is resident within the community to improve the quality of the decision-making process and in the preparation of plans. This principle honors and values the local knowledge and expertise that is found in many communities. It also develops and nurtures individuals with specialized expertise to become community resources after the current planning process is concluded and create a cadre of talented community volunteers. We encounter community experts in every neighborhood. Often, their expertise is not related to their work, but their passion or extracurricular interest. We have met experts who understand the city's water and sewer supply, those that have an encyclopedic knowledge about the buildings in their neighborhood, and amateur storytellers who can describe and document neighborhood change. Collectively, tapping this expertise allows planners to understand the values and ideals of a community as the engagement process unfolds.

6.4.4 Design an Inclusive Community Outreach Strategy

Crafting an inclusive community outreach strategy is challenging, and especially so, if there are budgetary constraints. At the beginning of this chapter, we discussed the concepts of social capital and the networks of trust and association that seem to be important in promoting social cohesion and a commitment to civic engagement. Planners can help to design opportunities for bridging social capital to emerge – by creating opportunities of people to work collaboratively with other people in their community whom they may have not have had a chance to encounter before. Increasing the opportunities to hear and learn from a diversity of voices is essential to a successful civic engagement strategy.

A. *Draw Study Area Boundaries*

As discussed earlier, the definition of a study area is an art. One technique we suggest is to overlay study boundaries from different projects that have happened in the area over time to see if there are any visible differences. Another technique is to ask long-term residents about the geographic extents of their neighborhood or community. Yet another technique is to use GIS and statistical software to understand how the demographic composition of the community can shift if a few census blocks are included or excluded in defining a study area. While there is no precise right or wrong answer, the delineation of a study area boundary is a political act, not a technical one.

B. *Identify Stakeholders*

As previously indicated, the community is not a monolithic whole. True, there is a sense of geographic community, particularly in neighborhoods that are well established over time. There are also communities of interest that may crisscross the study area. Identifying stakeholders requires more than listing the most obvious racial and ethnic groups to identify new and interesting ways in which the community can be encouraged to engage. Relying on formal, established stakeholder groups can be a starting point. In areas with small populations, individuals representing sectoral interests can be assembled to form a group that represents a particular geography.

C. *Develop Outreach Plan*

The purpose of developing an outreach plan is to find ways to connect with different stakeholder groups, in other words, connect with multiple publics. We recommend that planners consider connection points in person (face to face) and online across place and time.

Same place – same time Example: Open houses	*Same places – different times* Example: Community-led walking tours
Different places – same time Example: Electronic town hall	*Different places – different times* Example: Online surveys

Community conversations are at the heart of any outreach endeavor and they are one of the most conventional, reliable, and one of the most rewarding activities associated with community engagement. Holding conversations in real time can be done throughout the course of a project, but they are particularly important in the early stages. It is important to identify those individuals in the community who are communicators and nodes in information networks. It may be a small business owner, the head of the local parent teacher association (PTA), or the editor of a community newsletter. In any event, begin talking to them to get the word out that a community engagement process is beginning and that they should get involved. In all probability, they will advise you on who else the planning team should reach out to and provide appropriate contact information.

Focus groups are a specialized kind of structured community conversation. They are an ideal way to get a group together to clarify high-priority issues of concern in the community or to solicit feedback about how to approach a topic. With a good facilitator, an hour-long focused conversation can provide insights about how small groups of people think and feel about a planning problem. Focus groups can be scheduled in advance and organized with a minimum of fuss. Focus groups work very well when the participants have similar interests or are part of the same interest group – for example, conducting a focus group of small business owners who all provide retail services to the community may work better than trying to manage a focus group that includes both retailers and restaurateurs. This is because retail businesses have different kinds of needs and challenges than restaurants and other food service establishments. Thus, putting together focus groups must be approached carefully and systematically to avoid confusion.

Town hall meetings are another type of community conversation. They can offer great benefits because they create opportunities for a robust and broad dialogue with a large swatch of the community in a short period of time, say in a 2-h evening meeting. Organizing and logistics are slightly complicated – most communities will require them to be scheduled several months in advance. Depending on the scale of the project, planners can use a town hall meeting between one and three times over the course of a year-long project. The town hall meeting agenda has to be carefully planned to ensure that the process complies fully with the principles of community engagement described in the previous section.

Community-led walking/biking tours are a great way to engage active community leaders to showcase their skills and expertise. Having a resident or business owner lead a tour of a neighborhood is a wonderful way to understand the values and places that townspeople hold dear. Some of these activities are better suited for good weather, but these activities can provide additional context to support archival sources like previous planning studies or newspaper reports.

Conversations occur in space and place

Open houses, in the context of this discussion, indicate an accessible physical location that can be made available for walk-in appointments for the public to drop in to discuss a variety of issues and concerns over the life of a project. Vacant storefronts can be repurposed for this task, for example, with a commitment to be available for a couple of hours every week. Once a pattern is established and it is understood that someone will be available, members of the public can feel motivated to visit and engage in the project or projects that are being discussed. Artifacts from work in progress can be presented for review and discussion. Likewise, a computer with an internet connection can allow participants to look at different information online, take an online survey, or otherwise engage in some collaborative activity while they go about their routines – a stop on Saturday morning at a community open house is a nice break from shopping or running errands. By going to where the people already are, planners increase opportunities for meaningful community engagement.

Strategy: Host an Open House

If you are a planner assigned to introduce a project to a community, consider hosting an open house in partnership with a local business, locally based nonprofit organization, or a house of worship. In an open house, planners make themselves available to answer questions about their project to drop-in visitors. In addition, planners can create a series of activities to engage neighborhood residents in discussing community concerns.

Hosting an open house requires some preplanning and organization. The location of the open house must be familiar and accessible to people in the neighborhood. Open houses that are scheduled after regularly scheduled activities work well because they accommodate the pace of activities in the community and don't feel like an imposition. Inviting participants to attend and reminding them is another key to having a well-attended open house event. An open house can be distinguished and differentiated from a conventional community meeting because neighbors can drop in and stay for as little or for as long as they want.

It is beneficial to plan several activities to engage different demographic groups – consider different activities for children, teenagers, and elderly, all with the goal of being able to better understand the strengths/assets of the community as well to gather some preliminary data about problems that can be addressed through a planning process. The open house is also an opportunity for a planner to describe the scope and extent of the planning project and document community concerns (if any) about the project.

Every engagement process should be an opportunity to educate citizens about planning methods and techniques, not through lectures and exams but through informal activities. Young people can be invited to develop their own futuristic visions through art projects, while adults could engage in researching precedents or ideas that they would like to see implemented in their community. Static information displays with contact information can be left in local establishments such as cafes and other hangout spots. Community members should be tasked with the responsibilities of setting up these informational displays – their conversations with business owners will expand the reach of the project and demonstrate that different members of the community are invested in this process.

It is important for planners to recognize the value and benefits of online community engagement as part of a community outreach effort
Digital technologies allow for completely asynchronous communication. A website with downloadable information, online surveys, polls, participatory mapping activities, and other interactive digital methods discussed in Chapter 4 can be developed and used throughout the process. Millennials prefer to participate online rather than face to face. Live streaming of town hall meetings will allow young people and elderly people who are not able to easily travel to join in a meeting can gather a lot of information if they can watch the proceedings via a computer with an internet connection. Smaller groups can conduct a great deal of productive work via conference calls and Web-enabled meetings where documents can be viewed (in shared mode) and discussed. Both free and paid services to facilitate these conversations are now available, and every planning team should take full advantage of them. A webinar is an elegant digital alternative to a town hall meeting, especially if it is intended to have one-way information sharing with a limited question and answer session. People can participate in informal polls or raise questions in this format. The entire webinar can be archived and made available for nonparticipants who may want to access the information later.

D. *Establish Timeline and Reporting Milestones*

An outreach strategy must be calibrated to fit the scope of the project and the expectations of the community. Yet, all outreach strategies must include a clear timeline, including measurable milestones, an estimate of the approximate number of people who will be "touched" through the outreach, and a clarity about how each stakeholder group will be contacted. This document requires detail and clarity, and vague notations should be avoided.

E. *Publish and Communicate Outreach Plan*

Once the plan is finished, the general outline (including timeline and milestones) should be reported out to stakeholder groups, the community at large, and within the planning agency. Outreach is perhaps a misnomer here, signifying uni-directional communication. It should be noted that we have made every effort to create and foster dialogue throughout.

6.5 Closing Comments About Engagement

The planning team should develop an engagement strategy that is suited for the scale of the project and should maximize community-based resources. Securing these resources is also an important way to engage the community – in-kind contributions from local businesses, time offered by volunteers, and interest from local experts in getting involved are all measures of a successful engagement strategy. The planning team must identify community leaders who are not always leading projects but those who are new and eager to get involved and help prepare them to take on leadership roles such as facilitating small group conversations, staffing open houses, and playing a role in managing the outreach process in some way.

Although the engagement strategy may unfold organically, a successful engagement strategy should be planned with a good understanding of the needs of the project at hand and the community's own resources and tolerances for community engagement. Some communities are process oriented and are interested in how decisions are being made, while others are focused on seeing tangible outcomes of the decisions and knowing that their participation resulted in specific accomplishments. In most situations, communities vacillate between the two approaches and a successful engagement process must offer satisfaction in both ways. An engagement strategy's cadences and milestones should be meaningful to participants. This is only possible if the planner understands her community. Scheduling an important community meeting on a religious holiday or holding a town hall meeting during public school vacation are embarrassing and avoidable gaffes that nevertheless occur with depressing regularity.

6.6 Social Media

No planning project can be successful without a social media strategy. Social media can be used for both for information dissemination, encouraging peer-to-peer information sharing, and for increasing visibility about a project. Social media platforms are discussed in detail in Chapter 7. There is a variety of risks associated with the use of social media. One of the main concerns that are often raised by legal experts and risk-averse individuals is that the message can get lost and, worse, get co-opted and misused in counterproductive ways. In addition, government agencies have additional restrictions in using social media platforms because they cannot be seen lobbying for policy outcomes. Despite such fears, it is imperative to develop and roll out a social media strategy to support meaningful civic engagement. Initially, it is useful to start small, selecting one platform, such as *Twitter* to disseminate information as it happens and provide links to a static digital website so that the average onlooker can keep themselves informed about the project as it evolves. It is useful to develop some appropriate hashtags that can be used to consolidate the discussions that occur about the project.

6.7 Conclusions

In this chapter, we have made several points. Chief among them is to avoid falling into a trap of an adversarial strategy that creates winners and losers. Planning agencies are easily co-opted and at the same time vilified by savvy community groups who understand the power of 1960s style organizing and protest politics. Moving planning away from a game where political favors are given or withheld toward a more collaborative model of civic engagement requires time and patience. At the end of every engagement process, planners should ask themselves the following questions:

- Is there more willingness to participate in the future?
- Do more people trust the process, even when they disagree about the outcomes?
- Do newcomers who joined the process better understand how planning works, its strengths and its limits?
- Have participants become better at articulating in sharing their personal and local knowledge for the benefit of the community?
- Has the social, intellectual, political capacity of the participants improved?
- Is the process sustainable without planners?

If the answer to these questions is affirmative, then that is a measure of a good civic engagement process. Ultimately, engaged participants understand how to transform spaces and places using data, information, historical precedents, and with a realistic understanding of global and regional challenges. Planners are vital in facilitating civic engagement because they can educate, inform, and support the public in articulating the strengths of a community and shape their wishes and ideas into tangible projects and programs.

References

Arnstein S (1969) A ladder of citizen participation. J Am Inst Plann 8(3):216–224

Bebbington A, Woolcock M, Guggenheim S, Olson E (eds) (2006) The search for empowerment: social capital as idea and practice at the World Bank. Kumarian Press, West Hartford

Creighton J (2005) The public participation handbook: making better decisions through citizen involvement. Jossey-Bass, San Francisco

Davidoff P (1965) Advocacy and pluralism in planning. J Am Inst Plann 31(4):331–338. https://doi.org/10.1080/01944366508978187

Hall P (1996) Cities of tomorrow: an intellectual history of urban planning and design in the twentieth century., Updated edn. Blackwell, Malden

Jacobs J (1961) The death and life of great American cities. Random House, New York

Putnam R (2000) Bowling alone: the collapse and revival of American community. Simon & Schuster, New York

Chapter 7
Implementation

7.1 Moving from Ideas to Action

Recent planning graduates and new entrants to the planning profession are often overwhelmed about how best to be effective in their job. Most planning professionals enter the field with high-minded values such as social justice and equity and subsequently struggle to identify ways to "make a difference" within the confines of their role within an organization. Entry-level planners, especially those working in complex and hierarchically organized bureaucracies are assigned specialized tasks that often preclude them from understanding the big picture. Even in organizations that have a relatively flat managerial structure, planners may not be familiar with the myriad of ways in which their individual contributions interact with the work of others within and outside the organization. Nowhere is this disconnect more obvious than when a project moves from the creative planning phase to the pragmatic operational or implementation phase. This chapter describes the dynamics of implementation. We argue that a planner who understands these dynamics is better prepared to intervene to secure the integrity of ideas that they have championed in the planning phase. Only then, will they be able to help achieve project/policy implementation – an achievement that all planners celebrate. Successful implementation of plans often provides individual planners and planning teams a sense of accomplishment and job satisfaction.

What is implementation? Implementation usually references a wide range of activities or actions that occur after a decision has been made – decisions that include a requirement of action. Implementation is shaped by the scale, context, technological, and sociopolitical variables. The disconnect between plan making and implementation failures is a perpetual challenge; it casts negative aspersions on the value of planning as an approach to solving social problems in a democratic society.

Most planners would agree that the completion of a plan is an important milestone and an achievement to be celebrated. For instance, for transportation planners and supporters of public transit advocating for reducing auto dependence, the

L. Ramasubramanian, J. Albrecht, *Essential Methods for Planning Practitioners*, The Urban Book Series, https://doi.org/10.1007/978-3-319-68041-5_7

unveiling of a new city-wide greenway plan that encourages walking and bicycling is a memorable achievement because it is a product, an artifact that synthesizes the outcomes of their efforts which includes a complex array of data gathering and analysis tasks, as well as a series of community consultations to shore up support for the plan. Yet, no planner wants the plan they worked on to become part of a library's archive and never see the light of day again. Plans, as important and significant as they are, are often not celebrated unless the ideas contained within them are enacted to shape the built environment. The development of policies and programs encounters the same challenges – their value is full realized only when they are adopted within the corpus of existing laws and regulations that govern everyday life.

Planning professionals serve in multiple roles to translate ideas into actions. Their contributions to plan making are supported by a variety of *research* and *analysis* methods discussed in Chapters 4 and 5. The methods and techniques of *civic engagement* necessary to garner stakeholders' support for planning actions are discussed in Chapter 6. This chapter discusses the skills, methods, and techniques that are necessary in the long implementation and evaluation phases of a project, program, or policy.

7.2 Understanding Implementation Cycles

Implementation is challenging, even in the case of well-defined plans that are modest in scope, completed on time and under budget, and where the implementation falls within the purview of a single agency. When a project, policy, or plan is implemented, its immediate impact can sometimes be observed and measured immediately. For example, a transportation planner working on designing or planning a new bridge can celebrate the completion and opening of that bridge. They can also assess the effectiveness of the new bridge by assessing (through empirical methods) whether the new bridge serves its stated purpose – achieving a reduction in traffic congestion.

The impacts of many social policies, plans, and programs are difficult to verify immediately because implementation usually occurs in phases. Even in the case of the bridge example, there may be second-order effects that occur over time. For example, traffic flows could increase, thereby increasing congestion, rather than reducing it. Another second-order effect is that the new bridge could spur new development initiatives. Thus, both positive and negative effects of planning interventions can be difficult to assess and quantify immediately. Returning to the earlier example of greenway planning, planners and other greenway enthusiasts will not be able to accurately measure, quantify, or qualify all the societal benefits of the greenway immediately after the first phase is completed.

The complexity of implementation cycles can best be understood and explained by examining the history of large-scale infrastructure or planning projects. Alan Altschuler and David Luberoff (2003) define *mega*projects as "initiatives that are physical, very expensive, and public" and state that they are a "fundamentally an expression of public authority" (p.2). Among the many projects that qualify for this categorization are Boston's Central Artery/Tunnel (CA/T) project and the Tappan Zee Bridge replacement project in New York.

The New NY Bridge project, now known as the Mario Cuomo Bridge, currently underway (scheduled completion in 2018), is intended to replace the 3.1 mile Tappan Zee Bridge originally constructed in 1955. The project's estimated cost of USD 3.98 billion and the length of time from when the first discussions about bridge replacement began in 1997 to when construction began in late 2013 qualify it to be acknowledged as a planning megaproject. Philip Plotch (2015) describes this torturous path using the mnemonic, FAILURE, to explain the different factors that delayed project implementation. The factors include insufficient *f*unding, *a*dverse goals, *i*nteragency conflict, *l*ack of leadership, *u*ncertainty about alternatives, *r*egulations that were onerous, and *e*xpectations that were unrealistic (p.182). Each of these factors provides insight into the complexities associated with plan implementation. Long and costly delays are not exclusively linked to complex megaprojects. This book does not focus on the reasons (or factors in Plotch's parlance) typically associated with implementation delays; instead, we describe some valuable strategies and tactics that planners can adopt to minimize delays and manage stakeholders' expectations before and during implementation.

The skills that planners require to move ideas into tangible outcomes undoubtedly begin with plan making. But a good plan does not automatically result in an implemented plan! Most agency archives include a variety of plans that were never implemented. We argue that in the current social and political environment, planners, more than ever before, must be involved in conversations about implementation regardless of their area of specialization or expertise. As educators, we understand that graduate students and new planners aspire to become specialists in a specific area of expertise because it conveys job preparedness and readiness. We do not minimize this desire to cultivate specialized skills. However, planners cannot disengage from conversations about implementation.

One of the difficulties in moving from ideas to action is associated with reminding different stakeholders about the scope, purpose, and intended contributions of a project, plan, or policy. In linear time, stakeholders engaged in early stages may shift their focus or disengage completely by the time the plans are completed. Those who got involved to champion a single cause or issue are most likely to disengage once their ideas are considered, but the average member of the public has neither the inclination nor the stamina to be associated with the project for its entire planning and implementation life cycle. Citizens also trust that planners will implement the wishes that were expressed in public meetings. Having fulfilled their civic obligations, they move on to deal with their own lives. There is a collective amnesia that surrounds many long-term projects – goals change, scope creeps occur, and in a dynamic system, changes to parts of the system diminish the value of the project or plan that is being implemented. For example, a mobilized community's efforts to create local, well-paying jobs can be derailed by a national recession, an event over which they have no control. A thoughtful planner serving as a community memory keeper can play an important role in supporting the implementation of good projects, plans, and ideas.

In Chapter 2, we discussed some of the process-oriented challenges of planning. These included addressing the distrust of the public, the notion of "planning expertise," navigating the complexity of planning with and for a diverse populous, and finally, recognizing that the ways in which citizens' groups, bureaucrats, and elected officials engage with each other and with professional planners have changed over the years. Also in Chapter 2, we reminded planners that planning is both an art and a science, and requires the innovative use of available tools, including digital technologies. In the next section, we discuss the concept of *digital storytelling* to support implementation.

7.3 Digital Storytelling

Stories, or narratives, play a critically important role in communicating our perspective within our community and to the broader public. An effectively told story can influence decision-makers and spark wider discussions in our cities and neighborhoods (Schön and Rein 1994; Throgmorton 1996; Forester 1999). Some planning scholars have advocated for the integration of storytelling techniques to persuade and convince stakeholders about the value of specific planning decisions. Within this framework, the narrator does not merely provide the data and information necessary to make the decision – they use their skills to craft a situated and contextualized narrative that allows decision-makers to fully understand the significance of the decisions they are about to make. Critics point out that the storytelling is just that, stories – that relying on storytelling can sometimes mean succumbing to potential misrepresentations and falsehoods that are ultimately harmful within the context of planning. Throgmorton, an advocate of storytelling, also summarizes the critique of planning scholars like Flyvbjerg (1998) who observe that all the persuasive and eloquent storytelling in the world cannot outweigh power differentials, in other words, "power, not stories, is what matters" (Throgmorton 2003:127).

We use the phrase *digital storytelling* to describe socio-spatial narratives created by combining a variety of digital techniques. The use of this genre to inform and educate was popularized by documentary film makers like Ken Burns who combined still photographs, narration, actors reading quotes, as well as diaries and interviews in the 1990 TV series "The Civil War" to create a very comprehensive yet accessible collage of the military, social, and political facets and dynamics of that period in the history of the USA. Yet, Ken Burns has an advantage over the average planner – he tells a story about the past, whereas planners are called upon to speak persuasively about the future.

We are all experts about where we live, and our stories about places we know and love (or know and want to change) can be an evocative, eloquent, and compelling force for change through urban planning. Individuals or interest groups compile their understanding of a place and its history using photographs, video, animation, sound, music, text, and/or a narrative voice. For example, creating an audio narrative of citizens speaking about the positives and negatives of their neighborhood is more powerful than a planners' reporting of the same information that was derived from census data. More importantly, personal narratives add new data – they provide planners and decision-makers with *new* information which simply did not exist before! Nevertheless, census data is incredibly valuable. In our work with community groups, we have taken our data to the community for members to pore over it and confirm or disconfirm our work. Community residents are excited when the census data confirms what they know "in their gut." At the same time, the data sometimes challenges the cherished assumptions and biases of residents in a neighborhood. Planners as skilled facilitators (see Sect. 7.8) can help to explain these discrepancies. However, sometimes, even the best planners have to go back and check their numbers!

Experiential knowledge is also needed for further analyses – different types of data can be used for different purposes. In a community planning process, digital storytelling can be an effective way of gathering and sharing information on what is important to community members and what makes a community unique. By combining elements of public history and public art with storytelling, place construction can be redefined from current mainstream experience to include forgotten "invisible" parts of the city and make them visible.

Table 7.1 describes the strengths and limitations of digital storytelling. Planners or stakeholders can use digital storytelling techniques to describe a future vision or visions (scenarios), and it might also be used to illustrate the intended results of an agreed-upon plan. Narrative approaches can be related to formal modeling approaches (see Chapter 5). Narratives can be mined to provide the input data for urban models (Guhathakurta 2001, 2002). Narratives and quantitative models can be combined to develop scenarios that can be run at workshops with a variety of stakeholders. Narrative-driven scenarios are also an important component of "community visioning," which is a strategy used by urban planners to encourage the building of sustainable futures for communities through civic engagement (Ding 2005). As with any activity, it requires careful planning to make sure the stories will effectively inform a planning process.

Table 7.1 Strengths and limitations of digital storytelling

Strengths	Limitations
People can describe communities, places, and memories in their own words	Requires specific technologies and technical expertise
The narratives can be edited and adapted for different audiences	Digital technologies constantly change and evolve; keeping up may be challenging
Different points of view can be expressed/juxtaposed	Like many participatory tools, storytelling (digital or not) is limited by the number and quality of the participants
Effective way to engage youth and audiences without formal planning experience	Hard to introduce to groups that are comfortable with traditional engagement techniques
Effective way to create intergenerational and intercultural conversations	Developing the storyline and story production can be time-consuming
Many free and user-friendly tools to get started with digital storytelling	Poorly produced digital narratives can introduce bias that is not representative
Narratives can be archived and shared to create a community memory	

Digital storytelling activities are effective when they engage a broad diversity of participants and when they are integrated as a part of community conversations between and among different publics. Storytelling allows everyday citizens to effectively participate in value-mapping activities because the stories help elucidate and clarify the values contained within quantitative information. For example, the Orton Foundation's Community Heart & Soul model integrates digital storytelling into its phased approach to community planning with its three core principles: (1) involving everyone, (2) focusing on what matters, and (3) playing the long game. This model showcases the seamless integration of technology and storytelling to craft a narrative that advocates for the preservation of small towns and rural communities.[1] Although these narratives can be used during any stage in a planning process, for example, in a data-gathering (analysis) phase, we present this technique in Chapter 7 because we argue that it is very powerful toward the end of a planning project, when planners move from a phase of data analysis to data synthesis and implementation.

7.4 Understanding the Governance Landscape

Since a clear majority of entry-level planners will work for the public sector either directly or through their work in private consulting firms that fulfill planning functions for government agencies, we urge planning students and new graduates to invest time understanding the functions of different state and local agencies to understand where different types of planning occur. In most American cities, many traditional planning functions like zoning and land-use controls reside at the local (city/county) level. In a small- or medium-sized city, the planning department is an integral part of the city's management team, and the planning director has a great deal of power and influence about how important social and political concerns affecting that city are addressed. However, economic development, infrastructure provision and maintenance, environmental protection, and many other aspects of our public life are the responsibility of many government agencies that cross jurisdictional boundaries. Spatial extents and high density are but two of the variables that can add complexity.

In larger cities, public authorities (corporate instruments that undertake bureaucratic obligations and have broad powers and autonomy) and public-private partnerships P3s (contractual agreements between public agencies and private sector entities that deliver services or create facilities used by the public at large) routinely engage in major planning activities. In addition, a variety of nongovernmental organizations that include organized advocacy groups and informal issue-based coalitions and individual stakeholders also influence planning.

Recent graduates, and others new to the field, should learn about the diversity of planning activities that are undertaken within city and state government. One way to

[1] Orton Foundation's Community Heart & Soul model. Interactive website available at https://www.orton.org/build-your-community/model/, retrieved on April 2, 2017.

accomplish this goal is to understand the different types of work that agencies undertake. Planning activities can be examined across sectors, such as housing, economic development, education, and transportation. They can also be examined across a range of sociodemographic issues, chief among them inclusion, community development, aging populations, and youth services. The protection of the natural environment, energy production and conservation, and water and waste management also involve a great deal of planning. In short, planning occurs just about everywhere. It's the responsibility of serious planners or aspiring planning professionals to understand the diversity and complexity of planning from an agency perspective. For new graduates, this scan of agencies will provide valuable information about places to look for future employment.

Planning approaches may vary drastically from agency to agency. Kelly and Becker (2000) describe the different approaches to community-wide planning. They include goal-driven planning (a classic and traditional approach), trends-driven planning (a technocratic and incremental strategy that assumes that the present uses and activities will continue at a greater or lesser intensity), opportunity-driven planning (planning based on assessing needs using techniques such as SWOT analysis to determine intervention strategies), issue-driven planning (a pragmatic and results oriented, with expectations of immediate wins), and vision-driven planning (a principled long-term approach that can help create a culture of planning within a community). We would like to include crisis-driven planning (short-term planning to address immediate concerns, e.g., planning for the resettlement of refugees or addressing the after effects of a natural disaster) to this list.

In some cities, organized nongovernmental groups wield great influence in shaping conversations about planning. These include civic and advocacy organizations that champion specific planning approaches and ideas. In the New York region, the Regional Plan Association (http://www.rpa.org/) advocates its visions through the development of regional plans. Since the 1920s, the organization has published three plans, and some of the ideas contained within have been adopted and adapted by planning agencies in the region. At present the organization is working on the Fourth Regional Plan. The Congress for the New Urbanism (https://www.cnu.org/), founded in the early 1990s, has been influential over the last two decades in encouraging elected officials, architects, landscape architects, and planners to work collaboratively to mitigate urban sprawl through innovations in design and planning. New Urbanism includes a set of planning theories and frameworks and is coupled with a social movement that organizes and engages citizens in planning. If planners remember their planning history (see Chapter 2), then, they will not and should not be surprised that New Urbanist ideals have gained social and political currency.

Research institutes also influence public debates about planning. Some research institutes are more politicized than others, but all such institutes and think tanks use data to inform, educate, persuade, and ultimately focus decision-making. Examples of such influential institutions that have shaped planning discourse include the Robert Wood Johnson Foundation (https://www.rwjf.org/), the Brookings Institution (https://www.brookings.edu/), and the Urban Institute (https://www.urban.org/) among others.

To the extent where public participation is encouraged, community-based activists also strive to have their say. Organized activism continues to shape and limit the actions of many local and state governments. Last but not the least, individual citizens, John and Jane Q Public, also offer inputs directly, unfiltered by intermediaries as they have always done – through direct participation in planning processes, by sharing their views in public forums, and through the ballot box – electing those who are likely to share their views. In short, planning activities occur within and outside of the formally established departments of planning.

7.5 Understanding Policymaking Frameworks

What is public policymaking? A basic definition comes from Thomas Dye, "Public Policy is whatever governments choose to do or not to do" (Dye 1972:1). While simple, it seems to be rather vague. Policy appears to be very similar to planning – in that many scholars agree that the word *policy* describes the general principles that help to guide action. A simple way to separate policy from planning, for our purposes, is to consider public policy to be "principles that are made explicit in law and other *formal acts of governmental bodies*" (Preston and Post 2013: 11, emphasis added). The policy sciences literature, however, cautions us that many policies in practice are implicit, in that they "can be implemented without formal articulation of individual actions and decisions" (Preston and Post 2013:11). Policymaking can also be programmatic, different from those principles that are enacted as laws. Planners working at the national (consider federal agencies) or supranational scale (consider the United Nations or the World Bank) are more likely to be involved directly in policy analysis and public administration than those who work for local governments.

Policies can vary in scope and influence. For example, a government agency or department can design, implement, and enforce policies to achieve its legally mandated obligations. These policies tend to define the power and influence of the agency making the policy and may not impact a wide swath of society. There are three main models of policymaking – optimization, incrementalism, and power and bargaining. The fourth, institutional systems model integrates all three and is prevalent in complex democratic societies (Preston and Post 2013).

Policymaking and planning can seem virtually indistinguishable when we examine the rational-comprehensive approach to planning, popular after World War II. Within this framework, planners made technical and scientific information available to decision-makers in the political arena. They were charged to find optimally feasible solutions for well-defined problems, based on criteria articulated by decision-makers. This is the optimization model of policymaking (Preston and Post 2013).

Incrementalism popularized by Charles Lindblom (1959) argues that policies are made gradually, in stages. This model begins with the premise that decision-makers are not in agreement about overarching goals or the criteria that will help arrive at an optimal solution. Consider the Tappan Zee Bridge project we discussed earlier in this chapter. The reasons for the delays are attributable easily to the lack of agreement about goals and the key criteria that were needed to be satisfied to arrive at an optimal solution. If history is to be our guide, then most policymaking falls under the incremental model. Within the incremental planning model of policymaking, decision-makers develop policies to address immediate and near future concerns. Policies become a scaffold built to support specific ideas being championed and are more likely to address procedural concerns than resolving ideological conflicts.

The power and bargaining model of policymaking has its roots in political science. This model focuses on interactions between different societal groups, recognizing that there are many different constituencies that have varying levels of social, economic, and political power to influence the actions of government and that individuals can simultaneously be members of different constituencies. In its simplest form, a bargain requires mutually acceptable divisions of rewards among the groups that are involved in the negotiations. Accurately understanding the relative power of various groups that are involved and devising good communication processes becomes critical before one can successfully advocate for or against a specific policy position (Preston and Post 2013). This model of policymaking, anchored as it is in power, does not privilege technical expertise or normative positions – relying entirely on the positions that are articulated by those with power and influence over important decisions. Advocacy planning, discussed in Chapter 6, can be viewed as an off-shoot of, or as a reaction to this model of policymaking.

In contemporary societies, these three models of policymaking often coexist and operate alongside each other. The institutional systems model best describes their integration. Here, policymaking begins with formulations that are articulated and imbued with constitutional and governmental authority. Groups in our society raise issues that constitutional and governmental systems are required to address; these groups also simultaneously react to the actions taken by these systems. Once issues are brought to the fore, the optimization model of policymaking proceeds in an incremental fashion, with a recognition of the power differentials that exist among groups. Institutional systems analysis recognizes that the implicit norms and values that govern policymaking can change over time (Preston and Post 2013). In this policymaking landscape, planners must recognize where, when, and how to act to gain support and credence for their ideas. One way to identify those moments and points of inflection is to understand the nature of agenda setting in policymaking.

7.6 Agenda Setting and the Role of Social Media

In describing the institutional systems analysis approach to policymaking, we previously observed that groups in society exert pressure on governments to enact laws, policies, and rules and they also react and respond to the laws and policies enacted by governments. Planners must ask, "How do governments decide which issues to take on?" An ancillary question to consider is, "Why now?" Let us look outside the world of planning to understand this dynamic better. In the United States, the Supreme Court is the highest court of law; the cases the court decides to hear are widely accepted as important public policy issues because they have become worthy of the court's consideration. The decision to accept or reject a case for consideration is not taken lightly by the judges – and there are many procedural guidelines and steps that are followed. Yet, some issues that once were not on the Supreme Court's radar eventually become worthy of their consideration. How does that happen? As individuals, we hold opinions or perspectives about a range of issues that are relevant in the public realm. Opinions matter – sometimes, the same opinion is held by many people (for example, regarding the use of eminent domain authority); at other times, many people recognize that even if an issue affects only a subset of the population (e.g., transgender individuals), its meaningful resolution concerns society in its entirety.

Public opinion is an important way in which issues get placed on the court's agenda. It's reasonable to conclude that the results of opinion polls matter to policymakers. Opinion polls are conducted and developed by many organizations and the media. Despite credible research and practical evidence that suggests that opinion polls can be unreliable, they do have the power to create news headlines and thereby exert their own influence on policy conversations.

Interest groups have always worked to shape public opinion. Setting aside the interest groups that work in the arena of electoral politics, there are many interest groups that are issue-based coalitions that work to create visibility for specific issues and actively help to shape the policy to address their concerns. The American Planning Association in 2014 released the results of their national polling – unsurprisingly, they found that 67% of those surveyed believed that "community planning is important for economic recovery" (American Planning Association 2014). While the report offers a rich source of data and insights about how different demographic groups in the United States think about planning and the value of planning, one could not be faulted for wondering whether this was a self-serving exercise.

The American Planning Association (APA) also regularly assembles the collective thinking about issues that are important to planners. The policy guides[2] cover many different topics and include aging in community, agricultural land preservation, billboard controls, climate change, community residences, endangered species and habitat protection, energy, environment-waste management, environment-wetlands, factory-built housing, community and regional food planning, freight, hazard mitigation, historic and cultural resources, homelessness, housing, impact fees, neighborhood collaborative planning, provision of childcare, public redevelopment, security, smart growth, surface transportation, takings, and water. The association notes that three new topics – planning and health, affordable housing, and social equity and inclusive growth – have been approved for study and policy guide development.

Individual leaders often use their expertise to shape policy conversations. In New York City, *William H Whyte*, a sociologist, became interested in how people

[2]American Planning Association APA Policy Guides. Available at https://www.planning.org/policy/guides/, retrieved on April 14, 2017.

move in public spaces. His landmark book and accompanying documentary, *The Social Life of Small Urban Spaces* (1980), used time-lapse photography, observations, behavioral maps, and movie footage of public spaces to challenge many existing planning policies. His work was influential in creating physical changes (the transformation of Bryant Park in Manhattan) which is one notable example, but it also facilitated a whole series of policy changes about the design and management of urban public spaces. These policies were adopted by the New York Department of City Planning and continue to influence urban design in the city to this day. Another example of an influential leader shaping urban design is the architect *Jan Gehl*. Jan Gehl designed interventions that created pedestrian friendly public spaces. Cities throughout the world have used Gehl's work to develop policy and design guidelines to create pedestrian- and bike-friendly cities.

While the *media* has always helped to shape the policy agenda (there is a strong relationship between government officials and elected officials on the one hand and journalists on the other), the phenomenon of social media, including but not limited to texting, blogging, instagramming, and live streaming, and other individualized and hyper-local journalistic endeavors must be taken seriously by planners. In this century, *social media* is a significant influencer of public policy, escalating issues quickly to the realm of national debate and discussion. Sometimes, these high-energy viral transmissions of information are like electrical voltage fluctuations – they are an anomaly that is recorded but one that can ultimately be ignored. Yet, at other times, social media dialogues and hashtag activism serve as a bell weather that reflects changing public sentiments about social and political issues that directly impact how public policy issues are framed and reframed.

7.7 A Brief Comment About Budgets

Planners should understand how the budget for the agency they work for is structured and whether the budget allocations have increased or declined or stayed constant over the years. After adjusting for inflation, the budget is a reliable indicator about the government's overall policy priorities. A budget also shapes the procedures an agency can use to allocate resources for public programs and services, including how the performance of those programs and services will be evaluated. While outside of the scope of this book, we strongly recommend that planners make every effort to understand the basic principles of budgeting. An agency's budget usually includes an operating budget (expenditures for the current year) and a capital budget (a plan for long-term development or maintenance of new facilities or equipment). Open and/or publicly available data are useful to understand city and agency budgets, spending patterns, and how taxpayer dollars are allocated and distributed across different types of government functions. In sum, and to put it bluntly, *budgets strongly define the success and failure of policy priorities*.

While the structural opportunities and constraints that shape project implementation can seem daunting, we suggest that planners can continue to grow their sphere of influence over public policymaking within and outside their agencies by cultivating a range of skills discussed in the next chapter. You may wonder why a book focused on planning methods takes the time to discuss the development of personal and interpersonal skills. Our experience leads us to understand that successful planners possess a unique blend of technical, social, and political skills that enable them to shape and facilitate conversations about social and political change.

7.8 Ethics

As professionals, planners are expected to conform to ethical guidelines in their practice, regardless of whether they work for the government, private, or nonprofit sector. Acting ethically can be interpreted differently – depending on the situational context. For example, advocacy planners may differ from planners working for government on how to interpret situational ethics, but they would both agree that accepting bribes is *verboten*.

The key to acting ethically is to be well informed about the situational context and to act responsibly and thoughtfully. There are two sets of ethical guidelines that can provide guidance to all planners, especially newcomers to the field. The American Planning Association is a membership organization that welcomes members who are interested in planning regardless of their organizational affiliation. The American Institute of Certified Planners is an association of credentialed planners who demonstrate education, experience, and knowledge of the profession (qualifying for and passing the AICP exam) and commit to stay current on planning issues (Certification Maintenance). Both organizations have ethics guidelines. There is much agreement between the APA's statement of ethical principles in planning which is a guide and the AICP Code of Ethics; there are also some differences that may shape how individual planners holding an AICP certification address situational conflicts.

7.9 Conclusion

Planning professionals should consider the complex governance regimes and policy frameworks that drive implementation in the localities where they work. In addition to having a good understanding of technical skills and civic engagement methods, planning professionals should cultivate a variety of skills to assist everyday people in sifting through different kinds of information, ensuring that diverse voices are heard, and ultimately forging consensus about how to address shared community concerns about the future. When planners distance themselves from these obligations, confining themselves to technocratic roles, they diminish the influence of the profession to create meaningful social transformation.

References

Altschuler A, Luberoff D (2003) Mega-projects: the changing politics of urban public investment. Brookings Institution Press, Washington, DC

American Planning Association (2014) Investing in place for economic growth and competitiveness: a research summary, May 2014. American Planning Association, Chicago

Ding P (2005) Envisioning local futures: the evolution of community visioning as a tool for managing change. J Fut Stud 9(4):89–100

Dye T (1972) Understanding public policy. Prentice Hall, Inc, Englewood Cliffs

Flyvbjerg B (1998) Rationality and power: democracy in practice. University of Chicago Press, Chicago

Forester J (1999) The deliberative practitioner: encouraging participatory planning processes. The MIT Press, Cambridge, MA

Guhathakurta S (2001) Urban modeling as storytelling: using simulation as a narrative. The Bartlett Centre for Advanced Spatial Analysis (CASA) working paper series, 37. Available through CASA

Guhathakurta S (2002) Urban modeling as storytelling: using simulation models as a narrative. Environ Plan B Plan Des 29(6):895–911

Kelly E, Becker B (2000) Community planning: an introduction to the comprehensive plan. Island Press, Washington, DC

Lindblom C (1959) The science of muddling through. Public Adm Rev 19(2):79–99

Plotch P (2015) Politics across the Hudson: the Tappan Zee MegaProject. Rutgers University Press, New Brunswick

Preston L, Post J (2013) Private management and public policy: the principle of public responsibility. Stanford University Press, Stanford

Schön M, Rein M (1994) Frame reflection: toward the resolution of intractable policy controversies. Basic Books, Boston

Throgmorton J (1996) Planning as persuasive storytelling: the rhetorical construction of Chicago's electric future. University of Chicago Press, Chicago

Throgmorton J (2003) Planning as persuasive storytelling in a global-scale web of relationships. Plan Theory 2(2):125–151

Whyte W (1980) The social life of small urban spaces. Project for Public Spaces, New York. (published 2001)

Chapter 8
Epilogue

8.1 Introduction

We are nearing the end of our narrative. Our narrative recognizes the powerful role that planners can play in creating transformational change. The most obvious contributions that planners make reside in their ability to shape the built environment. This is important and should not be overlooked. Safe streets, well-planned neighborhoods with amenities that support the needs of residents and visitors, and welcoming public spaces support economic development and contribute to an enhanced quality of life. Yet, planners can and indeed do much more! Planners identify myriad opportunities for intervention by developing policies, programs, plans, and actions, all designed to address the needs of communities, while working within existing institutional frameworks and competing political agendas to address the needs and aspirations of the communities they serve.

According to the American Planning Association, a sizable majority of planners (70%) work in the public sector.[1] Among these public sector planners, about 40% are employed in city governments. For planners working in government, their work and self-expression are shaped and circumscribed by the politics, policies, attitudes, and beliefs of elected officials and appointed agency heads. Simply put, planners working in government are not able to articulate or implement ambitious planning agendas. They opt to use incremental planning approaches, as discussed in Chapter 7.

This is no excuse for a new planner to retreat into a world governed by technocratic expertise. The need to engage with political actors should be welcomed, rather than feared. We anticipate that planners reading this book will consider

[1] The 2016 APA/AICP Planners Salary Survey Employment Characteristics. The 2016 web-based survey was conducted by Readex Research for APA and AICP. The survey was sent to APA regular, life, faculty, and new professional members (28,856). The survey achieved an overall response rate of 39 percent. The results for salary are reported for 9709 respondents who provide salary data and indicated that they are employed or self-employed full time and year-round as planners or in planning-related positions. Data available from the American Planning Association.

L. Ramasubramanian, J. Albrecht, *Essential Methods for Planning Practitioners*,
The Urban Book Series, https://doi.org/10.1007/978-3-319-68041-5_8

working within and through formal institutions to create transformational change. All planners, especially those working in government, would do well to cultivate a wide range of planning skills – it is imperative to creating and then growing one's individual sphere of influence.

8.2 Planning Skills

As we have observed in Chapters 1 and 2, planners are interested in future-oriented problem solving. In a democracy, planners are (or should be) at the front lines, mediating and negotiating between competing interests and helping to forge a strong and hopeful vision for a shared future. To be successful at their job, they must understand the community they serve in all its complexity. Chapters 3, 4, and 5 provide some guidance about how planners can immerse themselves in understanding their community using a variety of data sources. That immersion is a wonderful first step and requires time and dedication to do correctly. They must also understand how to think critically about the built environment and its impact on the lived experience (Talen 2009).

8.2.1 Written Communication Skills

Planners must be able to communicate their knowledge and understanding of issues and methods to a range of nontechnical actors, chief among them citizen planners and elected officials who may be called upon to make legally binding decisions based on the data, analyses, and recommendations provided to them by planners who are expected to have some specialized expertise. These types of written communication can include staff reports, environment impact statements, and advisory memos. Although some of these documents can be quite lengthy and follow prescribed formats, a typical planning report or memo is relatively short. Most memos include the description of the problem or planning issue, an analysis of alternatives, and a set of recommendations as suggested actions. The format and content are highly variable depending on the needs of the agency.

Bonnie Johnson and Ward Lyles (2016) conducted a systematic analysis of staff reports that examined noncontroversial development applications, building a representative national sample of 94 reports, one each from 94 cities across 41 states. They concluded:

> Many staff reports provide traditional basic information, but do not summarize that information or use modern formatting tools to make text more comprehensible. Most staff reports reference the comprehensive plan but rarely cite the future landuse map or plans of adjacent jurisdictions. Many mention checking traditional public facilities like roads, but rarely list parks or pedestrian or cycling facilities. Most do not include maps, arguments for recommendations, or references to soliciting public input. (Johnson and Lyles 2016; 22)

Their findings reveal the challenges of writing for nontechnical audiences in a persuasive and reliable way without compromising their technical integrity. This balancing act – of making practical judgements, based on available data and evidence – is a skill that must be cultivated. When planners provide jargon-filled technical narratives that stop short of drawing inferences and meaningful recommendations, they are not living up to the ideals of the field and the profession. Even though much of the writing within the world of planning can be labeled technical writing, it must inform, engage, and spur action.

8.2.2 Graphical Communication Skills

There is an extensive literature in planning that speaks to the power of hand-drawn sketches (e.g., Alexander et al. 1977; Lynch 1960), maps (McHarg 1969), and other 2D and 3D renderings, models, and simulations (e.g., Brail and Klosterman 2001) to inform and educate the public. Practicing planners must understand the technical skills of mapping and visualizing data using Geographic Information Science (GIS) tools (Albrecht 2007). See Chapter 3 for the use of these tools and techniques to describe a community in all its richness and complexity. Planning documents, reports, and memos should integrate graphics to communicate concepts that are more easily understood in graphical rather than in narrative formats. Michael Kwartler and Gianni Longo (2008) argue that visioning allows for greater transparency and inclusiveness and provides participants with a comprehensive understanding of the issues being discussed. Furthermore, proponents of 3D visualizations and realistic simulations argue that the lay public (and more importantly elected officials) can better understand the physical contours of a proposed project, before breaking ground. While this is true, professional planners must take responsibility for grounding the soaring creativity embedded in artistic renderings to match lived and budgetary realities (not every plan warrants a Geodesign solution). Thus, planners must familiarize themselves with the strengths and limits of graphical visualization techniques.

Learning the use of graphical tools and techniques is a demanding but ultimately rewarding endeavor. For new planners, staying current with new graphics and visualization technologies is an imperative. Fortunately, there are many nonproprietary and free-to-use tools which are now available, and this list continues to grow. Keeping up with evolving technologies can sometimes seem a Sisyphean task. Notably, tech meetups[2] allow planners to interact with their peers as well as tool developers. A plethora of online resources like webinars can help planners stay abreast of evolving tech trends.

[2] https://www.meetup.com/.

8.2.3 Public Presentation Skills

Planners are often called upon to make presentations to different audiences. Planners working in the public interest must present information and evidence to the lay public in a respectful way. Being prepared to make public presentations and being an engaging speaker make planners accessible to community members. People who attend community meetings are more likely to ask questions and participate in planning activities if the planner attends community meetings and events on a regular basis. In this way, planners are akin to politicians – they must be visible and accessible to the communities they serve. New planners should take every opportunity to attend community meetings and observe senior planners and local elected officials in action. Ultimately, each planner cultivates his or her own style of presenting information and engaging in public conversations, but here we end by observing that it is an important aspect of a successful planner's repertoire.

8.2.4 Consensus Building Skills

In engaging with the public, planners must be genuine and transparent brokers between different interest groups. They cannot "take sides" and must provide data and information equitably to interest groups. Securing agreement from all participants about the parameters of the process and its intended outcomes is a critical first step. Documenting and sharing points of agreement are also valuable. In community meetings, planners must remember that consensus does not mean complete agreement; it means that everyone is satisfied with the process and can abide by the results.

It is important to recognize that planners are often engaged simultaneously in substantive and procedural conversations – in other words, "what we are trying to do?" along with, "how are we trying to do it?" In these instances, the consensus building requires planners to pay attention to diverse stakeholders' interests as well as the power differences inherent in any social group. Planners may consider developing skills in designing and managing charrettes (see end of Sect. 8.3).

8.2.5 Social Communication Skills

Aspiring planners, and recent entrants to the planning profession, should develop a regimen for consuming and digesting social media narratives about the urban environments and sectors they work in. It may also be necessary to cultivate a social media presence and avidly manage and curate their presence to fully participate in these newly emergent forums for public policy conversations. We suggest caution though; not all conversations are civil in these online spaces, and the propensity for harassment is high. Planners should try to understand diverse policy perspectives. Nowadays, online forums make it possible for consumers to custom-tailor their information consumption, picking and choosing information that conforms to their world views and preconceived ideas. When carried to an extreme, many negative consequences can result – particularly troubling for planners who are expected to engage and serve all people, not just people holding ideologies that mirror their own.

8.3 Planning in the Twenty-First Century

Most planning textbooks outline a typical planning process. The description is usually accompanied by a diagram that includes a hierarchical series of inter-connected steps, moving from issue identification, goal formulation, data collection and analysis, development of alternatives, selection of a preferred alternative, implementation, and monitoring. We suggest that this formalized planning process is a twentieth-century ideal, better abandoned in theory and practice.

The case studies we describe in Chapter 3 provide a sketch of how a community can be described to facilitate a *center-out planning process* that integrates elements of top-down and bottom-up planning, acknowledging that planning in contemporary societies is initiated and supported by governments, civil society organizations, and the private sector. Our case studies emphasize and draw attention to the importance of constructing study area boundaries – boundaries matter a great deal. Even when our case study locations, Roosevelt Island and Hunts Point appear to be clearly delineated and stand out as distinct entities, closer scrutiny reveals a series of connections and dependencies with the rest of the city, for example, in the way political jurisdictions are drawn to include Roosevelt Island as a part of Manhattan for some planning purposes but not for others. A similar recognition of interdependency can be observed as we observe flows of people to and from Hunts Point for the purposes of employment. Planners should guard against study area boundaries that are drawn to avoid or ignore serious problems or contentious issues. A study that skirts around serious problems, even when well executed, will eventually prove to be useless.

Boundaries are not merely physical – planners should also be watchful when some issues are moved "off the table" because they do not fall neatly into a designated aspect of planning. Issues such as social inclusion (or the lack thereof) of any societal group must be addressed within the context of planning – it should not be set aside as someone else's problem to solve.

In both our case study examples, we have emphasized the importance of understanding history from the perspective of the people – in other words, to bolster official historical narratives with indigenous "people's narratives" of their place. In addition to validating and celebrating local experiential knowledge, understanding the people's history also prepares planners to identify fault lines of disagreement that can occur when envisioning the future.

Planners must use GIS and other visualization tools to develop a geographical history of a place – in other words, understand the interconnection between nature and the built environment. Twentieth-century planners have emphasized the idea of a *tabula rasa*, a clean slate upon which planners could imprint their singular visions (e.g., Peattie 1987). We state unequivocally that there is no such tabula rasa and planners must acknowledge this clearly. We began our narrative, referencing climate change and the importance about communicating complex information to a skeptical public. One of the ways to break the barriers of communication is to use geographical history – in Hunts Point, for example, the old maps distinguish between high ground and swamp land, and they show the rivers and the tributaries that once flowed to drain the landscape but have yielded to impervious paved surfaces. Hunts Point is not alone. Eric Sanderson's Welikia project,[3] for example, allows citizens to go back in time to better understand how their neighborhood must have been before humans settled there for the first time. It is these types of educational conversations that will allow planners to talk critically about contemporary planning challenges and opportunities.

Authors such as Kretzman and McKnight (1993) have pioneered an asset-building approach to planning, arguing that successful planning must identify and celebrate the unique assets of any community for that plan to create positive outcomes. This is true in recognizing the physical infrastructure assets as well as the cultural and social assets of any community. Engaging nontraditional planning participants, including children (e.g., Race and Torma 1998), can offer great rewards in understanding the rich tapestry of any physical setting.

Twenty-first-century planning cannot be reactive; it cannot be limited to quality-of-life planning, nor can it be limited to the creation of a laundry list of social issues waiting for someone else to solve them. If planners want to make a difference, they must learn to improve the quality of the civic engagement processes. We have discussed civic engagement throughout the book, beginning with specific methods of data collection in Chapters 4 and 5, the theories and conceptual ways to manage engagement in Chapter 6, and the integrative approaches of digital story telling in Chapter 7. Planning charrettes are one way to cultivate and nurture civic engagement in a neighborhood or community. Charrettes are useful to move ideas from planning to implementation, by developing specific ways for the public to think about the feasibility of visionary ideas. In other words, a charrette is a one-stop shop where experts from different disciplines work collaboratively with members of the public to create a workable plan (e.g., Lennertz and Lutzenhiser 2006; Condon 2008). Charrettes are an important element in a planning process. This process begins with planners and community members learning about the community together and developing a shared understanding of planning issues and problems and culminates with the implementation of an agreed-upon plan.

[3] Welikia project. Available at https://welikia.org/explore/mannahatta-map/. Accessed May 15, 2017.

8.4 Twenty-First-Century Planners

Over 25% of all planners working for the public sector work in areas with a population of half a million people or more.[4] Most planners work in cities or city-like environments, with only 5% of planners reporting that they work in a rural area according to the APA/AICP survey referenced earlier. These statistics should give the reader some pause. The 2016 survey also indicates that over half of the planners surveyed in 2016 were engaged in community development (53%). Land-use or code enforcement and transportation planning also attracted many planning professionals. It may be useful if more planners worked or chose to work in other less-developed planning specializations including housing, sustainability, facilities and infrastructure planning, participation and empowerment, spatial planning, and planning law.

Earlier, in Chapter 2, we argued that the three crosscutting planning challenges were (1) urbanization, (2) demography, and (3) climate change. The processes of urbanization hollow out the hinterland, creating new planning challenges and opportunities (e.g., Vance 2016). A recent essay in the *Wall Street Journal* noted that rural America is the new inner city.[5] Setting aside the fact that the reference promotes unhealthy stereotypes about the "inner city," the article observes that the people "left behind" in rural America tend to be poorer, unhealthier, and more collectively disadvantaged than their urban counterparts. In this context, we challenge planners, particularly those planners working on issues of community development to pay attention to the development of rural areas and small towns – this may call for newer and more innovative policies and practices of land management, education, health-care delivery, and workforce development. Likewise, demographic shifts impact the design of civic engagement processes and communication protocols – planning approaches that work effectively in a youthful community of highly educated and wealthy millennials may not work well in a community of new immigrants or older adults.

Finally, we return to the issue of climate change. National and international climate change policies have been crafted,[6] and most nations agree in principle about the need to change their patterns of consumption and production in order to reduce the deleterious effects of climate change. Cities and local governments throughout the world can and should play an important role in mitigating the harmful effects of climate change and helping citizens adapt to the changing climate.[7,8] Planners should be at the forefront of assisting with both the development of mitigation strategies and adaption planning by working in partnership with scientists, educators, and concerned publics to address these challenges.

---

[4] The 2016 APA/AICP Planners Salary Survey Employment Characteristics.

[5] Adamy, J. & Overburg, P. 2017. One Nation, Divisible | Rural America Is the New 'Inner City', The Wall Street Journal, May 26th, 2017. Available at https://www.wsj.com/articles/rural-america-is-the-new-inner-city-1495817008. Accessed on May 26, 2017.

[6] The United Nations Framework Convention on Climate Change. Available at http://unfccc.int/2860.php. Accessed on May 26, 2017.

[7] 100 Resilient Cities. Available at http://www.100resilientcities.org. Accessed on May 26, 2017.

[8] ICLEI-Local Governments for Sustainability. Available at http://www.iclei.org/. Accessed on May 27, 2017.

Scholars have the luxury of offering critique of planners who are data-driven or not data-driven enough, for instance, or about planning processes that wasted time on community consultation or were not consultative enough. Practising planners are criticized for focusing exclusively on quality-of-life issues and not taking on larger structural and societal challenges. Planners are criticized for not focusing on the needs of the middle class or for focusing entirely on them. So, it goes. As educators and practitioners, ourselves, we urge you, the reader, to listen and reflect on feedback and criticism but to not become paralyzed rendering yourself inactive or ineffective. Professional planners have the obligation to plan for everyone, even those who reject the need for planning. Planners are obligated to act ethically and responsibly to consider the needs of present and future generations – we propose that all twenty-first-century planners must make a serious commitment to planning for sustainability within their own area of specialization.

We believe that if planners want to shape and influence planning policies, one simple way to begin is to become engaged in shaping the policy agenda of the field's most prominent advocate, to join and participate in the membership association that represents the nation's planners. Another way to influence the agenda is by being engaged directly in community-based, or better, community-driven planning activities. Participating in community activities outside of the job is an important way to gain the trust and respect of the communities you serve.

Planners are storytellers, they analyze and synthesize, and, above all, they help to make sense of the present and the future. This is an important role and one that planners should not abdicate to others. We hope the methods and techniques we have discussed in this book encourage you to become better at your craft – doing planning is hard work, and the best planners make it look easy. You can tell stories with data and information, with maps and graphics, and with innovative ways to engage the communities you work with and work to bridge and resolve differences through conversations – conversations are at the heart of good planning.

We wrote this book for new planning graduates, planners in the early stages of their career, and planners who are making career transitions. We firmly believe that twenty-first-century planners should not shy away from learning innovative analytical methods and techniques. At the same time, we want to ensure that planning methods are situated and used appropriately within a social and political context, and with respectful engagement with multiple publics. We recognize that we are asking a lot of future planners – we do so because the field and the profession demand it, and we believe planners are more than up to taking on these challenges.

References

Albrecht J (2007) Key concepts and techniques in GIS. Sage, London

Alexander C, Ishikawa S, Silverstein M (1977) A pattern language: towns, buildings, construction. Oxford University Press, New York

Brail R, Klosterman RE (eds) (2001) Planning support systems: integrating geographic information systems, models, and visualization tools. ESRI Press, Redlands

Condon P (2008) Design charrettes for sustainable communities. Island Press, Washington, DC

Johnson B, Lyles W (2016) The unexamined staff report: results from an evaluation of a national sample. J Am Plan Assoc 82(1):22–36

Kretzman J, McKnight J (1993) Building communities from the inside out: a path to finding and mobilizing a community's assets. Center for Urban Affairs and Policy Research, Northwestern University, Evanston

Kwartler M, Longo G (2008) Visioning and Visualization: People, Pixels and Plans. Cambridge, MA: Lincoln Institute of Land Policy

Lennertz B, Lutzenhiser A (2006) The Charrette handbook: the essential guide for accelerated collaborative community planning. American Planning Association (Planners Press), Chicago

Lynch K (1960) The image of the city. MIT Press, Cambridge, MA

McHarg I (1969) Design with nature. Natural History Press, Garden City

Peattie L (1987) Planning: rethinking Ciudad Guyana. University of Michigan Press, Ann Arbor

Race B, Torma C (1998) Youth planning charrettes: a manual for planners, teachers, and youth advocates. American Planning Association (Planners Press), Chicago

Talen E (2009) Urban design reclaimed: tools, techniques, and strategies for planners. American Planning Association (Planners Press), Chicago

Vance JD (2016) Hillbilly elegy: a memoir of a family and culture in crisis. HarperCollins Publishers, New York

Index

L. Ramasubramanian, J. Albrecht, *Essential Methods for Planning Practitioners*, The Urban Book Series, https://doi.org/10.1007/978-3-319-68041-5

MIX
Papier aus verantwortungsvollen Quellen
Paper from responsible sources
FSC® C105338
FSC
www.fsc.org

If you have any concerns about our products,
you can contact us on
ProductSafety@springernature.com

In case Publisher is established outside the EU,
the EU authorized representative is:
Springer Nature Customer Service Center GmbH
Europaplatz 3, 69115 Heidelberg, Germany

Printed by Libri Plureos GmbH
in Hamburg, Germany